Recent Advances in Physiotherapy

WB 460 PAR   £31·99

Recent Advances in Physiotherapy

Edited by
CECILY PARTRIDGE

BICENTENNIAL
1807
WILEY
2007
BICENTENNIAL

John Wiley & Sons, Ltd

Other Wiley Editorial Offices

John Wiley & Sons Inc., 111 River Street, Hoboken, NJ 07030, USA

Jossey-Bass, 989 Market Street, San Francisco, CA 94103-1741, USA

Wiley-VCH Verlag GmbH, Boschstr. 12, D-69469 Weinheim, Germany

John Wiley & Sons Australia Ltd, 42 McDougall Street, Milton, Queensland 4064, Australia

John Wiley & Sons (Asia) Pte Ltd, 2 Clementi Loop #02-01, Jin Xing Distripark, Singapore 129809

John Wiley & Sons Canada Ltd, 6045 Freemont Blvd, Mississauga, ONT, L5R 4J3.

Wiley also publishes its books in a variety of electronic formats. Some content that appears in print may
not be available in electronic books.

Anniversary Logo Design: Richard J. Pacifico

Library of Congress Cataloging-in-Publication Data

Recent advances in physiotherapy / edited by Cecily Partridge.
    p. ; cm.
  ISBN-13: 978-0-470-02542-0
  ISBN-10: 0-470-02542-5
  1. Physical therapy.   I. Partridge, Cecily J.
  [DNLM: 1. Physical Therapy Modalities. WB 460 R295 2007]
  RM700.R36 2007
  615.8′2 – dc22
                2006032511

A catalogue record for this book is available from the British Library

ISBN-13: 978-0-470-02542-0

Typeset by TechBooks Electronic Services Pvt. Ltd., Delhi, India.
Printed and bound in Great Britain by TJ International Ltd, Padstow, Cornwall.

This book is printed on acid-free paper responsibly manufactured from sustainable forestry in which at
least two trees are planted for each one used for paper production.

Contents

Contributors

Cecily Partridge
PhD, BA Hons, FCSP
Cecily is an Honorary Reader in the Centre for Health Services Studies at the University of Kent at Canterbury and an Emeritus Reader of London University. Her research and clinical interests have been mainly in neurological physiotherapy and the use of appropriate research methods in physiotherapy. She set up the first UK MSc degree in Research Methods for therapists in 1980, and founded the journal *Physiotherapy Research International* in 1996 and was editor until 2006. *Centre for Health Service Studies, Cornwallis Building, The University of Kent, Canterbury, Kent CT2 7NF email: cecily.partridge@virgin.net*

Louise Ada
PhD, MA, BSc, Grad Dip Phty
Louise is an Associate Professor in the School of Physiotherapy, The University of Sydney. Her teaching and research are in the area of adult neurology. Her research covers: examining the contribution of motor impairments such as weakness, incoordination, spasticity to limitations in physical activity; testing interventions for stroke rehabilitation; and investigating the delivery of rehabilitation, in particular, increasing the amount of practice of physical activity. *School of Physiotherapy, Faculty of Health Sciences, The University of Sydney, PO Box 170, Lidcombe NSW 1825 Australia Fax: 61293519278 email: L.Ada@fhs.usyd.edu.au*

Laura Browning
BPhysio
Laura Browning graduated with a Bachelor of Physiotherapy from La Trobe University, Melbourne in 1999. She worked as a junior physiotherapist at the Western Hospital, Melbourne, before commencing as a cardiothoracic physiotherapist at the Royal Melbourne Hospital, while continuing her clinical role and teaching undergraduate cardiothoracic physiotherapy students at the university. Her research interests include functional recovery following abdominal surgery, post-operative mobilisation programmes, and physiotherapy practice in abdominal surgery.

Catherine M. Dean
PhD, MA, BAppSci (Phty)
Cath is senior lecturer in the School of Physiotherapy, The University of Sydney. Her teaching and research are in the areas of clinical education, adult neurology, and the older person. Her research covers: examining models of clinical education for physiotherapy students; testing interventions for stroke rehabilitation and the older person; and investigating the delivery of rehabilitation, in particular, increasing the

amount of practice of physical activity. *School of Physiotherapy, Faculty of Health Sciences, The University of Sydney, PO Box 170, Lidcombe NSW 1825 Australia Fax: 61293519278 email: C.Dean@fhs.usyd.edu.au*

Elizabeth Dean
PhD PT
Elizabeth Dean is professor on faculty in the School of Rehabilitation Sciences, University of British Columbia, Canada. She has been invited to speak worldwide. Because lifestyle conditions are no longer pandemic in western countries alone, her research has increasingly focused on integrating knowledge of culture and diversity in promoting health and wellness globally, and in addressing the physical therapy needs of people from the ICU to community. She has published widely and is a co-editor of the text 'Cardiovascular and Pulmonary Physical Therapy: Evidence and Practice (4 edn)'. She spent a year as Senior of the Cardiovascular/Cardiorespiratory Team, Kuwait Dalhousie Project, Kuwait, and a year as Visiting Professor at the Hong Kong Polytechnic University. *School of Rehabilitation Sciences, University of British Columbia, T325-2211Westbrook Mall, Vancouver, British Columbia, Canada V6T Fax: 16048227624 email: elizdean@interchange.ubc.ca*

Linda Denehy
PhD, BAppSc (Physio), Grad Dip Physio (Cardiothoracic)
Linda Denehy graduated as a physiotherapist in 1976 in Melbourne, and completed her Graduate Diploma of Physiotherapy (Cardiothoracic) in 1987 and her PhD in April 2001 at the University of Melbourne. She worked in major public hospitals in Melbourne for 15 years and at the Royal Brompton hospital in London for a year before pursuing an academic career. Linda is currently a senior lecturer in the School of Physiotherapy at the University of Melbourne, where she coordinates both the undergraduate and post-graduate cardiorespiratory programmes and supervises research higher degree students. Her primary research interests involve management of patients in the area of acute care, including major surgery and critical care. *Post-graduate Student Research Co-ordinator, School of Physiotherapy, Faculty of Medicine Dentistry and Health Sciences, University of Melbourne Victoria 3010 Australia email: l.denehy@unimelb.edu.au*

Katherine Durham
BSc (Hons)
Kathy Durham graduated as a physiotherapist in 1997 from Nottingham University. As a senior, she has worked within the fields of mental health, elderly rehabilitation and neurology. She has a broad background in the assessment and treatment of neurological conditions and has specialised in stroke rehabilitation. Kathy is currently working towards her doctorate at Birmingham University, looking at the effects of different types of feedback on motor performance.

Alison Harmer
PhD, BAppSc (Physio)
Alison Harmer is a lecturer in the School of Physiotherapy, The University of Sydney, Lidcombe, Australia. Alison has research interests in effects of exercise and exercise

training on muscle morphology and metabolism in patient populations, including those with diabetes, after joint replacement, and patients with back pain.

Lester Jones
MCSP
Lester Jones is a senior lecturer in the Faculty of Health and Social Care Sciences, Kingston University and St George's University of London. He has worked in interdisciplinary teams in rehabilitation (Royal Melbourne Hospital) and pain management (University of Sydney Pain Management and Research Centre) as well as in private practice (Sydney and London). He currently holds an honorary Senior Lecturer position in the Faculty of Medicine University of Sydney. Lester has completed Bachelor degrees in physiotherapy and psychology and a post-graduate diploma in behavioural studies in health care at La Trobe University, Melbourne; a post-graduate certificate in teaching and learning in higher education at Kingston University London; and a Master's degree in pain management at the University of Sydney. Lester has been on the committee of the Physiotherapy Pain Association (UK) for the last three years, two of those as Education Officer. He is also a member of the Australian Physiotherapy Association and the International Association for the study of pain, including the 'Pain and Movement' special interest group. *School of Physiotherapy, Faculty of Health and Social Care Science, 2nd Floor, Grosvenor Wing, St George's University of London, Cranmer Terrace London SW17 0RE, UK email: ljones@hscs.sgul.ac.uk*

Joy C. MacDermid
PhD, MSc, BS PT, BSc
Joy MacDermid is a physical therapist, hand therapist, epidemiologist, and holds a Canadian Institutes of Health Research (CIHR) New Investigator Award. She is an Associate Professor (School of Rehabilitation Science) at McMaster University and is also Co-director of the Clinical Research Lab within the Hand and Upper Limb Centre (HULC) in Canada. She is cross-appointed to Departments of Surgery and Epidemiology at both McMaster University and the University of Western Ontario. Her research interests include: upper extremity disability; randomized clinical trials and trial methodology; outcomes studies; psychometrics of clinical measurement (performance and self-report); clinical epidemiology; clinical practice guidelines; and knowledge transfer. Her research projects emphasise multidisciplinary approaches to enhancing prevention, assessment, and management of musculoskeletal problems. *School of Rehabilitation Science, McMaster University, 1400 Main Street West, Rm 429, IAHS, Hamilton, Ontario, L8S 1C7 Canada Phone: 9055259140 ext. 22524 Fax: 9055240069 email: macderj@mcmaster.ca*

Martine Nadler
PhD, MSc, MCSP
Martine Nadler qualified as a physiotherapist in 1987. She is a clinical specialist at the Wolfson Centre, Wimbledon, London (part of St George's Healthcare NHS Trust) and a part time post-graduate lecturer. In 1997, she read for a Master's in Neuroscience at the Department of Anatomy and Developmental Biology at University College London, and in 2000 published a PhD in the Department of Physiology at UCL.

In addition to working at various London teaching hospitals, she spent five years at the Bobath Centre London. Her research interests include investigation of central pathway changes after stroke. Dr Nadler currently holds an honorary research post at the Centre for Rehabilitation and Ageing at St George's Hospital, London. *115 Coombe Lane, Wimbledon, London SW20 OQY UK email: atgd80@dsl.pipex.com*

Justine Naylor
PhD, BAppSc (Physio)
Justine is Senior Research Fellow, Elective Orthopaedics, Sydney South West Area Health Service, NSW, Australia; Conjoint Senior Lecturer, UNSW; Honorary Fellow, University of Sydney. She has research interest in the fields of joint replacement surgery and cardiopulmonary physiotherapy. *Research and Quality Manager, Whitlam Joint Replacement Centre, Fairfield Hospital, New South Wales, Australia email: Justine.Naylor@swsahs.nsw.gov.au*

Jennifer A. Pryor
PhD, MBA, MSc, FNZSP, MCSP
Jennifer Pryor trained as a physiotherapist in New Zealand, but has worked for many years at Royal Brompton Hospital, London. She is currently the Senior Research Fellow in Physiotherapy at the Hospital, and an Honorary Lecturer at University College London. At University College she is involved with the MSc and Certificate Courses in Advanced Cardiorespiratory Physiotherapy. She is co-editor of the text-book *Physiotherapy for Respiratory and Cardiac Problems: adults and children* and her doctorate was on airway clearance in people with cystic fibrosis. She has many peer review publications and has lectured throughout Europe and in the United States, Brazil and New Zealand. *Royal Brompton Hospital, Sydney St., London SW3 6NP UK email: J.Pryor@rbh.nthames.nhs.uk*

Paulette M. Van Vliet
PhD, MSc, BAppSc (Physio)
Paulette is currently a research fellow at the School of Health Sciences at the University of Birmingham in the UK. She worked as a physiotherapist in neurological rehabilitation for ten years, before moving on to a career researching and lecturing on the subject. Her research interests are recovery of upper limb motor control after stroke; evaluation and development of physiotherapy intervention for stroke patients; and skill acquisition following stroke. Recent research has involved a randomised controlled trial comparing a Bobath-based and a Movement Science-based approach to stroke re-habilitation. Current research focuses on the temporal coordination of reach-to-grasp in patients with stroke, and the effects of different types of feedback on motor learning after stroke. She also lectures to post-graduate and undergraduate physiotherapy students on issues related to stroke rehabilitation. *School of Health Sciences, University of Birmingham, Edgbaston B15 2TT UK email: paulette.vanvliet@ntworld.com*

Richard Walker
MBBS, FRACS (Orth)
Orthopaedic Surgeon, Arthroplasty and Trauma Surgeon at Sydney Bone and Joint Clinic, VMO Liverpool Hospital, Sydney South West Area Health Service, NSW, Australia.

Nicola Walsh
MSc, MCSP
After gaining clinical experience in a variety of musculoskeletal settings, including professional sport and a diagnostic gait laboratory, Nicola was employed as a lecturer/practitioner at King's College London. She then worked as a research associate for four years on an Arthritis Research Campaign (ARC) funded randomised controlled trial (RCT) of a clinical cost-effective rehabilitation programme for chronic knee pain and osteoarthritis (OA) in primary care. This work forms part of her ongoing PhD (funded by the ARC) investigating long-term physiotherapy management strategies for lower limb osteoarthritis. In addition, she is lead investigator on a UK Physiotherapy Research Foundation RCT looking at an exercise and self management regimen for hip OA. Currently Nicola is employed as a senior lecturer at the University of the West of England. *Faculty of Health and Social Care, Glenside Campus, University of West England, Blackberry Hill, Bristol BS16 1DD UK email: Walsh@uwe.ac.uk*

Audrey Wang
MSNZS
Audrey Wang is a Clinical Specialist Physiotherapist at INPUT, Pain Management Unit, St Thomas' Hospital, London. Her experience includes working in interdisciplinary teams in chronic fatigue management (Essex Centre for Neurosciences) and pain management services, including return to work rehabilitation and case management in the United Kingdom. Her involvement in research projects includes the Job Retention and Rehabilitation Pilot (Work Care) – a Department of Work and Pensions and Department of Health initiative – and fatigue in primary care. She has also worked within the public and private sector in New Zealand. Having completed her Bachelor's degree in Physiotherapy at Otago University, Dunedin, New Zealand, she is presently undertaking her dissertation for her Master of Science in Applied Biomechanics with the University of Strathclyde, Glasgow. Audrey is also a member of the British Pain Society, Chartered Society of Physiotherapy (UK), and Physiotherapy Pain Association (UK).

Introduction

CECILY PARTRIDGE

The purpose of this book, the second in the series, is to enable those with an interest in physiotherapy to keep up to date with recent research relating to the profession, and in particular to provide information about the current bases of evidence for treatments frequently used for common conditions. The first book was restricted to the evidence-base for physiotherapy for neurological conditions; this one also deals with other conditions treated by physiotherapists, including respiratory, musculoskeletal, surgical, orthopaedic, post-operative and pain problems.

The book will be of interest to a wide range of physiotherapists, both undergraduate and post-graduate, to those who refer their patients for physiotherapy, and to administrators and others who commission physiotherapy services.

Each chapter starts with a Case Report of a real patient. This format was adopted to encourage dialogue between clinicians and researchers and stress the relevance of research to practice. A gap is often evident between the two but it is essential for the advancement of the profession that research both is, and is seen as, relevant to practice.

The authors of the chapters were selected as specialists in their own fields and as having both clinical and research expertise. Brief biographies are given to provide some idea of their very wide range of experience and specialisation. To ensure some consistency, authors were asked to follow the style of the previous book. Essentially they were asked to use the patient in the Case Report as a starting point to describe the treatment approaches they would prescribe, then to ask clinically relevant questions as a way of citing the current evidence-base for the treatment.

To enable the reader to estimate the strength of the evidence presented, authors were asked to rate the references they cited on the scale provided. This was adapted for physiotherapy from those first presented by Sackett et al. (2000). The original medical scales were not considered appropriate for physiotherapy because the randomised controlled trial (RCT) is widely regarded as the gold standard in medical research but has not yet been demonstrated as such in physiotherapy. In most evaluations of medical treatment there is a clear diagnosis ratified by clear criteria and usually supported by laboratory tests; the treatment can be administered in pre-specified doses; medication for the control group can be indistinguishable from the active preparation; and results can again be confirmed by the use of well validated tests. In many areas of physiotherapy the diagnosis is often unclear, as can be seen from the

Recent Advances in Physiotherapy. Edited by C. Partridge
© 2007 John Wiley & Sons, Ltd

case reports; interventions may be adapted to suit the individual, and cannot therefore be pre-specified; and in addition, outcomes are often behaviourally defined. These all mitigate against using the RCT as the gold standard in physiotherapy. Some authors also question its status in medicine. Goodman (1998, 1999) maintained that 'most RCTs are conducted on unrepresentative populations of heterogeneous patients and interpretation of results is usually far from straightforward'.

The three broad categories of the scale are given below:

A Based on the results of sound research, citing the results clearly, often a clinical trial, but to include single case study design. Also sound qualitative research, for example exploring patients' mood states or opinions.

B Laboratory based investigations in, for example, biomechanics, or neurophysiology where results help to inform practice but have not been evaluated in the treatment of patients.

C Statements provided by authority figures. Also citations from textbooks and consensus statements.

Reviews were marked as **R**. References without any letter did not fit any of these categories. Where, infrequently, unpublished PhDs were cited they were labelled as **A/R**.

The authors themselves assigned the reference categories using these criteria. Though the term 'evidence-based practice' is currently widely used, key to developing sound practice is the collaboration between researchers and clinicians to try to ensure researchers are tackling clinically relevant questions. One of the strengths of this book is the overt linking of practice to research, with authors having both research and clinical experience.

REFERENCES

Goodman NW (1998) Anaesthesia and evidence based medicine. *Anaesthesia* **53**: 353–68.
Goodman NW (1999) Who will challenge evidence based medicine? *Journal of the Royal College of Physicians* **33**: 249–51.
Sackett DL, Straus SE, Richardson WS, Rosenberg W, Haynes RB (2000) *Evidence Based Medicine. How to practice and teach EBM* (2 edn) Edinburgh: Churchill Livingstone, pp. 3–4.

I Cardiorespiratory

1 Physiotherapy and the Adult with Non-Cystic Fibrosis Bronchiectasis

JENNIFER A. PRYOR

INTRODUCTION

Bronchiectasis is defined as 'abnormal chronic dilatation of one or more bronchi' (Wilson 2003 **C**). The face of bronchiectasis is changing (Greenstone 2002 **C**). It used to be characterised by large volumes of purulent sputum, but today may also be characterised by a persistent and irritating non-productive cough. With the increasing use of antibiotics in the treatment of pulmonary infections in childhood, many patients with bronchiectasis have an underlying disease that predisposes them to chronic or recurrent infection, for example cystic fibrosis, immunodeficiency including HIV, primary ciliary dyskinesia, allergic bronchopulmonary aspergillosis and Mycobacterium avium complex (Rosen 2006 **C**). Diagnosis was by plain chest radiograph, with the extent of the disease assessed by bronchography (injection of contrast into the bronchial airway), but this was an invasive and unpleasant procedure. Today high-resolution computed tomography (thin slices taken through both lungs) allows identification of thickened bronchial walls, bronchial dilatation and ring opacities containing air-fluid levels (Copley et al. 2002 **C**) (see Figure 1.1).

This chapter will present two cases with diagnoses of bronchiectasis, referred for 'chest physiotherapy', one with severe bronchiectasis and one with mild bronchiectasis. Both patients had significant problems.

CASE REPORT I

Mrs AH, aged 58, presented with a chronic cough productive of copious amounts of purulent sputum and fatigue. Mrs AH's high-resolution computed tomography showed extensive bronchiectasis in both lower lobes associated with patchy consolidation and mucus plugging. The distribution was thought to be typical for a post-pertussis syndrome as the cause of her bronchiectasis. Her full lung function studies indicated severe airflow limitation with three-quarters of a litre of gas trapping and marked reduction in spirometric indices. Her gas transfer coefficient was 'reasonably' well preserved. End capillary carbon dioxide was at the upper limit of normal and there

Recent Advances in Physiotherapy. Edited by C. Partridge
© 2007 John Wiley & Sons, Ltd

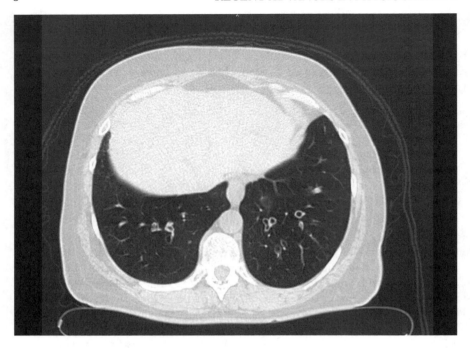

Figure 1.1. High-resolution computed tomography (CT) through the lower lobes, showing the classic signet ring sign (dilated bronchus with adjacent pulmonary artery of normal size) seen in established bronchiectasis.

was evidence of mild hypoxaemia. Haematological and biochemical indices showed mild microcytosis with no significant anaemia or abnormality in immunoglobulins. Her sputum cultured Pseudomonas aeruginosa. On auscultation there were coarse crackles throughout both lung fields.

Mrs AH's medical management included the introduction of an aggressive cyclical antibiotic regimen to reduce the bacterial load and an inhaled corticosteroid was introduced to suppress airway inflammation. She had received physiotherapy for her chest, in the form of airway clearance, in the Middle East. This had comprised the head-down tilt position with chest clapping from an assistant, and coughing when secretions reached the upper airways. The physiotherapist visited twice a week, no airway clearance was undertaken in between times and there was no encouragement to undertake a programme of physical exercise.

QUESTION 1

Which airway clearance regimen should be recommended for an adult with bronchiectasis?

A search for the evidence for airway clearance in bronchiectasis was undertaken in February 2006 using the key words 'physiotherapy' or 'physical therapy' and 'bronchiectasis'. This revealed nothing on the Cochrane database but using 'bronchiectasis' alone, two systematic reviews of interest were identified: 'Bronchopulmonary hygiene physical therapy for chronic obstructive pulmonary disease and bronchiectasis' (Jones & Rowe 2006 **R**) and 'Physical training for bronchiectasis' (Bradley et al. 2006 **R**). Jones and Rowe identified seven trials, which were said to be small and not generally of high quality. The authors said that in most comparisons, bronchial hygiene physical therapy produced no significant effects on pulmonary function, apart from clearing sputum. They concluded that there was not enough evidence to show whether there are benefits from chest physiotherapy to remove secretions from the lungs of people with chronic obstructive pulmonary disease or bronchiectasis.

The key word 'bronchiectasis' was used in the PEDro physiotherapy evidence database and identified 16 studies, 14 in English. Ten of these studies related to airway clearance and two to exercise. This database is one of the most efficient ways for the busy clinician to access some of the evidence, but not all clinical trials of relevance are included and it is therefore important to be aware of related publications in the field which can be accessed via Medline, Embase and the Cumulative Index to Nursing and Allied Health Literature (CINAHL). A systematic review requires evidence from randomised controlled trials and few have been undertaken in cardiorespiratory physiotherapy. This does not mean the evidence from other types of trial is invalid, but rather it means that valid 'low-grade' evidence, which may be of clinical significance, will probably not have been included in any systematic review.

The reviews on airway clearance do not address the physiological benefits of the removal of excess purulent secretions from the airways. Hypothetically, airway clearance techniques can decrease mucus plugging and aid in removing secretions containing inflammatory cells and by-products, thus decreasing damage to epithelia. In addition, movement and removal of bronchial secretions containing bacteria, especially Pseudomonas, may decrease local inflammatory responses and delay the change of Pseudomonas to mucoid morphology (Lapin C (2006) Personal communication **C**). Clinical expertise would support the practice of using an airway clearance technique in people with chronic sputum production and it is important to remember the definition of evidence by Sackett et al. (1996 **C**), that is, the integration of clinical expertise and the best available evidence from systematic research.

There are several airway clearance techniques which have been shown to aid the mobilisation and clearance of excess mucus from the airways. These include postural drainage and percussion (the regimen Mrs AH had been using, with assistance, in the Middle East) (Pryor et al. 1979 **A**), the active cycle of breathing techniques (Pryor et al. 1979 **A**; Thompson & Thompson 1968 **A**), autogenic drainage (Schöni 1989, **C**), positive expiratory pressure (Falk et al. 1984 **A**), oscillating positive expiratory pressure (Cegla et al. 1997 **A**; Konstan et al. 1994 **A**), high frequency chest wall oscillation (Warwick & Hansen 1991 **A**), intrapercussive pulmonary ventilation (Newhouse et al. 1998 **A**; Varekojis et al. 2003 **A**) and resistive inspiratory manoeuvres (Chatham et al. 2004 **A**; Patterson et al. 2004 **A**). Over 27 years ago, postural drainage and percussion was shown to be less effective than the active cycle of breathing techniques (Pryor et al. 1979 **A**) and yet it is still practised in many countries.

Many of the airway clearance studies have been undertaken in people with cystic fibrosis. Extrapolation to people with non-cystic fibrosis bronchiectasis must be with caution, but it is likely that the regimens of the active cycle of breathing techniques, autogenic drainage, positive expiratory pressure, oscillating positive expiratory pressure and high frequency chest wall oscillation are equally effective (Accurso et al. 2004 **A**; Patterson et al. 2005 **A**; Pryor 2005 **A**; Thompson et al. 2002 **A**). The choice of regimen may be one of personal preference, but this is likely to be influenced by the knowledge and experience of the physiotherapist. It is also likely that adherence to treatment will be increased if the airway clearance regimen is one which appeals to the patient and if they have been involved in the selection process. What is as yet unknown is whether a change of regimen, at intervals, will increase adherence to treatment.

Many countries use the sitting position for airway clearance. A study by Cecins et al. (1999 **A**), in people with bronchiectasis associated and not associated with cystic fibrosis, concluded that the side-lying position was as effective as the head-down tipped position and was preferred by the patients. Cystic fibrosis, in the early stages, is a disease which primarily affects the upper lobes bilaterally (Tomashefski et al. 1986 **B**). Bronchiectasis not associated with cystic fibrosis often presents with a middle and/or lower lobe distribution, indicative of a childhood viral infection. Generalised changes suggest an underlying host defence defect and an upper lobe unilateral problem, either post-tuberculosis or allergic bronchopulmonary aspergillosis (Greenstone 2002 **C**). The sitting position may be effective for people with cystic fibrosis, but this is not necessarily the best position for people with bronchiectasis not associated with cystic fibrosis and affecting the middle and/or lower lobes. In the individual patient, it is not difficult to solve this clinical problem. The patient should begin by using the selected airway clearance regimen in the sitting position. When the patient and the therapist have decided that continuing the treatment will not result in further expectoration of sputum, side lying with positioning for the affected segments should be tried. If more sputum is mobilised and cleared this will indicate there is benefit in using a side lying (lower lobes) or side lying $1/4$ turn from supine (middle zones) position.

Traditionally the emphasis for the use of gravity assisted positioning has been on the drainage of secretions (Ewart 1901 **C**). Wong et al. (1977 **A**), using radionuclide imaging techniques in patients with cystic fibrosis, demonstrated that an abnormal tracheal mucus clearance approached normal when the patients were placed in a 25 degree head-down tipped position. More recent work, using inhaled radiolabelled particles, found during postural drainage in people with cystic fibrosis that mucus clearance was greater from the dependent lung than from the uppermost lung (Lannefors & Wollmer 1992 **A**). This suggests that in mucus clearance the effect of the increase in regional lung ventilation may be greater than the direct effect of gravity.

An abscess cavity is likely to drain more effectively when the opening of the cavity points downwards, but today many people with bronchiectasis have only minor dilatation of the airway walls and the movement of mucus along these bronchiectatic airways may be better facilitated by the increase in airflow in the dependent lung than by the drainage effects of gravity in the uppermost lung, which were useful in the past. Theory would therefore indicate a patient with minimal right lower lobe bronchiectatic changes should be positioned in right side lying first, to increase ventilation, and then changed to left side lying.

Airflow is essential for airway clearance (Lapin 2002 **B**). There are similarities across most of the airway clearance regimens. All except autogenic drainage include the forced expiratory manoeuvre of huffing (Thompson & Thompson 1968 **A**), which increases expiratory flow, and this is now recognised as the most effective component of airway clearance (van der Schans 1997 **B**). Autogenic drainage utilises an unforced manoeuvre to augment expiratory flow (Schöni 1989 **B**), and the increase in expiratory flow of both the huff and an autogenic drainage breath should reduce the viscosity of mucus. This can be explained by its thixotropic property (Selsby & Jones 1990 **B**). The movement of secretions along the airways is said to be by either slug or annular flow (Lapin 2002 **B**; Selsby & Jones 1990 **B**). In addition, with the forced expiratory manoeuvre of the huff there is an oscillation of the airway walls (Freitag et al. 1989 **B**) which should further help to loosen secretions from them. Most of the regimens include a technique to increase lung volume and this is said to increase airflow via the collateral ventilatory channels (Macklem 1971 **B**), allowing air to flow behind secretions and to assist in mobilising them.

To return to Mrs AH, it was ethical to introduce an airway clearance regimen independent of an assistant to give her the opportunity to take responsibility for her management, and one which had been shown to be more effective than that of postural drainage and percussion. The two regimens not only independent of an assistant but also independent of a device are the active cycle of breathing techniques and autogenic drainage. The therapist's selection of one or other is probably influenced by their familiarity with the regimens.

For Mrs AH the active cycle of breathing techniques was chosen. The physiology behind the techniques of the active cycle of breathing was explained to Mrs AH. This included the loosening effect of the thoracic expansion exercises, utilising collateral ventilation to get the air in behind the mucus; the rest periods of breathing control; and the squeezing up of the excess bronchial secretions, from the choke points proximal to the equal pressure points, with huffing (the forced expiration technique (Pryor et al. 1979 **A**)). The techniques were practised with effect, initially in the sitting position and then in alternate side lying as the change in posture led to an increase in audible crackles from the airways. It was not long before Mrs AH developed an appreciation of how short or long a huff was required, dependent on the position of secretions within the airways, and a moderately copious amount of purulent secretions was expectorated. Mrs AH expressed her disappointment that she had not received any chest clapping and initially was not enthusiastic about continuing the regimen twice daily herself.

Self-chest clapping, in the stable clinical state, has not been shown to increase the expectoration of sputum (Webber et al. 1985 **A**). It could be argued that Mrs AH was not in a stable clinical state, but it was important to introduce a regimen which she could continue on her return to the Middle East and the introduction of self-chest clapping was likely to increase the work involved and detract from effective huffing.

Mrs AH returned for reassessment the following week. She had conscientiously undertaken the airway clearance regimen twice a day. Her sputum had decreased in purulence and quantity and she said that she was feeling much better and had more energy. The improvement is likely to have been owing to the combination of the medical management and adherence to an effective self-airway clearance regimen.

Additional techniques which may increase airway clearance in people with bronchiectasis include the nebulisation of normal saline and hypertonic saline (Kellett et al. 2005 A), humidification (Conway et al. 1992 A) and adrenoceptor agonists (Sutton et al. 1988 A). These, used together with airway clearance techniques, may enhance mucus clearance. Dornase alfa has not been shown to be of benefit in non-cystic fibrosis bronchiectasis and may lead to a reduction in lung function (Wills et al. 1996 A). Oral mucolytics, combined with antibiotics, may help sputum production and clearance (Crockett et al. 2006 A).

QUESTION 2

What is the evidence for physical training in an adult with bronchiectasis?

The fatigue experienced by Mrs AH is a characteristic of chronic chest infection and is usually associated with a decrease in exercise capacity together with increasing breathlessness on exertion, leading to a vicious cycle of increasing inactivity. Bradley, Moran and Greenstone (2006 R), in their systematic review on physical training for bronchiectasis, identified only two reports suggesting some benefits from inspiratory muscle training on exercise capacity, quality of life and respiratory muscle function. They concluded that further research is needed to assess the benefits of other types of physical training and pulmonary rehabilitation in bronchiectasis.

Much of the research in pulmonary rehabilitation has been in people with chronic obstructive pulmonary disease but people with bronchiectasis whose quality of life has been reduced by chronic breathlessness may also benefit (British Thoracic Society Standards of Care Subcommittee 2001 A). Newall et al. (2005 A), in people with bronchiectasis, compared pulmonary rehabilitation plus sham inspiratory muscle training, pulmonary rehabilitation with targeted inspiratory muscle training, and a control group with no intervention. They concluded that exercise training (pulmonary rehabilitation) improved exercise capacity in this group of patients and that inspiratory muscle training conferred no additional benefit.

Access to a full pulmonary rehabilitation programme is not always available and the vicious cycle of increasing inactivity can be broken by the simple progressive stair climbing programme designed by McGavin et al. (1977 A) and modified by Webber for use on the flat (Pryor 2004 C; Webber 1980 C). As Mrs AH was to return to her own country, which was different from that in which she was receiving treatment, the McGavin programme on the stairs was selected. The programme encourages the patient to exercise to breathless, in a defined and short period of time (eight weeks), with the understanding that breathlessness in this context is uncomfortable but not harmful. In between this daily exercise, breathlessness on exertion can be lessened by the introduction of breathing control (Rose 1999 **A**) to minimise the work of breathing. Positions which encourage the use of breathing control are said to be effective by altering the length tension status of the diaphragm, but the evidence is controversial (Gosselink et al. 1995 **A**) and it is important to assess and reassess the outcomes in the individual patient.

OUTCOME MEASUREMENTS

Outcome measurements for Case I could include: sputum volume or weight, sputum purulence (Miller 1963 **C**) (but sputum purulence is also likely to be affected by the antibiotic regimen), a field exercise test to measure exercise capacity (six-minute walking test (Butland et al. 1982 **A**) or shuttle walking test (Singh et al. 1992 **A**)) in association with a Borg scale (Borg 1982 **A**) of breathlessness and limb fatigue, and lung function.

CASE REPORT II

Mr SB, aged 30, presented with an irritating non-productive cough of 12 months, with each episode of coughing lasting for several minutes at a time, and being particularly troublesome at night on lying down. His partner had moved to a separate bedroom as she was unable to sleep with the persistent coughing. Stress, a change in air temperature and a change in posture could all precipitate bouts of coughing. Mr SB was a life-long non-smoker. There was no abnormality on his plain chest radiograph, and he had been given several courses of antibiotics and asthma management (British Thoracic Society & Scottish Intercollegiate Guidelines Network 2005), including inhaled sympathomimetic bronchodilators and inhaled corticosteroids, without effect. There was no evidence of a post-nasal drip or gastro-oesophageal reflux. He was finally referred to a specialist respiratory physician. High-resolution computed tomography revealed some changes in the right middle zone which just met the diagnostic criteria for bronchiectasis. His full lung function studies and gas transfer coefficient were all within the normal ranges. End capillary carbon dioxide was normal, and haematological and biochemical indices were normal with no immunoglobulin abnormality. His sputum culture was reported as 'No significant bacterial growth' and his chest was clear on auscultation, with normal breath sounds and no added sounds. The cause of his bronchiectasis was unknown, but may have been related to an episode of pneumonia in childhood. He was referred for physiotherapy.

QUESTION 1

Which is the evidence-based airway clearance regimen for an adult with bronchiectasis?

The literature search was as for Case I, but most of the subjects in the studies were expectorating sputum. Mr SB was not expectorating any sputum.

From previous clinical experience, the active cycle of breathing techniques was introduced with positioning for the right middle lobe. The first position was that of right side lying $\frac{1}{4}$ turn from supine to increase ventilation to the right middle zone. Mr SB's huff was initially dry sounding and non-productive, but with the breathing exercises it became moist sounding and Mr SB said that he could feel mucus coming up into the back of his throat, which he was aware of swallowing. The exercises were continued in left side lying $\frac{1}{4}$ turn from supine with similar results. The treatment time was about 15 minutes shared between the two positions, twice daily, and each session concluded with one or two huffs combined with breathing control in the sitting position.

Two days later, Mr SB was no longer complaining of a cough. The ongoing programme was a short daily check, in the sitting position, using the active cycle of breathing techniques. In the presence of any audible crackles on huffing, Mr SB was to progress to the side lying positions and to increase the time for treatment. He was also to follow this regimen if he thought he was getting, or if he developed, a chest infection. An alternative airway clearance regimen to that of the active cycle of breathing techniques could have been used dependent on the therapist's knowledge and expertise, and patient preference.

Using the forced expiration technique of the active cycle of breathing techniques, patients can be taught to recognise early crackles on huffing as a sign of excess mucus in the airways. The forced expiratory manoeuvre of huffing can be explained using the concept of the equal pressure point (West 1997 **B**). The equal pressure point (EPP) is the point where the pressure within the airway is equal to the pressure surrounding the airway. The airway downstream of the equal pressure point, towards the mouth, is compressed. This dynamic compression is an important mechanism which determines the efficacy of cough (Macklem 1974 **B**) and also applies to the forced expiratory manoeuvre of the huff. Proximal to the equal pressure point is the choke point (Dawson & Elliott 1977 **B**; Selsby & Jones 1990 **C**) and it is from this point, up towards the mouth, that there is a squeezing effect on the airway owing to the higher pressure outside the airway.

The positions of the equal pressure points are dependent on lung volume (West 1997 **B**). During normal tidal breathing and at a high lung volume, for example a spontaneous cough, the equal pressure points are said to be at the level of the carina or larger bronchi (Mead et al. 1967 **B**). As lung volume decreases, the equal pressure points move peripherally, allowing progressively deeper parts of the airways to be cleared. Without the need for a stethoscope, excess bronchial secretions produce audible coarse crackles during huffing. Crackles which occur with high lung volume huffing represent secretions in the larger proximal upper airways. If they occur with huffing at low lung volumes, secretions are likely to be in the smaller more peripheral airways and can be mobilised from bronchiectatic lung segments to non-bronchiectatic lung segments, where the normal mucociliary escalator should be effective in the cephalad movement of bronchial secretions.

Mr SB was not complaining of any increase in shortness of breath on exertion and was attending the gymnasium at his work place five days a week.

OUTCOME MEASUREMENTS

With computed tomography, bronchiectasis can be identified before the patient has developed a productive cough and the amount of sputum expectorated may not be an appropriate outcome measure for the effectiveness of treatment in these patients. Outcome measurements for Case II could include a visual analogue scale of cough or a valid and reliable cough-specific health-related quality of life instrument (Irwin et al. 2006 **A**).

COMMENT

The evidence and, in particular, systematic reviews alone are not yet able to answer many clinical questions in cardiorespiratory physiotherapy. The randomised controlled trial is not necessarily the best research methodology for clinical research questions in physiotherapy, but usually only research using the randomised controlled trial is considered for inclusion in systematic reviews. Recently the Cochrane Reviews have included the generic inverse variance method for meta-analysis of data from cross-over trials and data from parallel-designed trials, but even with these included the systematic review data for physiotherapy in bronchiectasis is limited.

Physiotherapy, rather than being 'evidence-based practice', should be 'practice-based evidence' (Lewis E (2004) Personal communication **C**), where the clinician generates the research questions for the researcher. This approach will lead more quickly to effective patient management and patient benefit. If the current approach to evidence-based practice, which has not itself been validated, is to continue, many physiotherapy techniques will be lost, not because they are ineffective but either because the randomised controlled trial has not been undertaken or because the right measurement tool has not been used or is not yet available. Future generations of physiotherapists must be very cautious in their interpretation of the evidence and take into consideration not only A grade evidence but also C grade evidence, of clinical experience and expertise.

REFERENCES

Accurso FJ, Sontag MK, Koenig JM, Quittner AL (2004) Multi-center airway secretion clearance study in cystic fibrosis. *Pediatric Pulmonology* Suppl. **27**: 314.

Borg GA (1982) Psychophysical bases of perceived exertion. *Medicine Science Sports Exercise* **14**(5): 377–381.

Bradley J, Moran F, Greenstone M (2006) Physical training for bronchiectasis (Review). *Cochrane Library* **2** http://www.thecochranelibrary.com.

British Thoracic Society, Scottish Intercollegiate Guidelines Network (2005) British Guideline on the management of asthma. http://www.sign.ac.uk/pdf/sign63.pdf.

British Thoracic Society Standards of Care Subcommittee (2001) Pulmonary rehabilitation. *Thorax* **56**(11): 827–834.

Butland RJ, Pang J, Gross ER, Woodcock AA, Geddes DM (1982) Two–, six–, and 12–minute walking tests in respiratory disease. *British Medical Journal* **284**(6329): 1607–1608.

Cecins NM, Jenkins SC, Pengelley J, Ryan G (1999) The active cycle of breathing techniques – to tip or not to tip? *Respiratory Medicine* **93**(9): 660–665.

Cegla UH, Bautz M, Fröde G, Werner T (1997) Physical therapy in patients with COAD and tracheobronchial instability – a comparison of two oscillating PEP systems (RC-Cornet®, VRP1 Desitin). Results of a randomised prospective study of 90 patients. *Pneumologie* **51**(2): 129–136.

Chatham K, Ionescu AA, Nixon LS, Shale DJ (2004) A short-term comparison of two methods of sputum expectoration in cystic fibrosis. *European Respiratory Journal* **23**(3): 435–439.

Conway JH, Fleming JS, Perring S, Holgate ST (1992) Humidification as an adjunct to chest physiotherapy in aiding tracheo-bronchial clearance in patients with bronchiectasis. *Respiratory Medicine* **86**(2): 109–114.

Copley SJ, Collins CD, Hansell DM (2002) Thoracic Imaging – adults. In: Pryor JA, Prasad (eds) *Physiotherapy for respiratory and cardiac problems* (3 edn) Edinburgh: Churchill Livingstone, pp. 27–53.

Crockett AJ, Cranston JM, Latimer KM, Alpers JH (2006) Mucolytics for bronchiectasis (Review). *Cochrane Library* **2** http://www.thecochranelibrary.com.

Dawson SV, Elliott EA (1977) Wave-speed limitation on expiratory flow – a unifying concept. *Journal of Applied Physiology* **43**(3): 498–515.

Ewart W (1901) The treatment of bronchiectasis and of chronic bronchial affections by posture and by respiratory exercises. *Lancet* **2**: 70–72.

Falk M, Kelstrup M, Andersen JB, Kinoshita T, Falk P, Støvring S, Gøthgen I (1984) Improving the ketchup bottle method with positive expiratory pressure, PEP, in cystic fibrosis. *European Journal of Respiratory Diseases* **65**(6): 423–432.

Freitag L, Bremme J, Schroer M (1989) High frequency oscillation for respiratory physiotherapy. *British Journal of Anaesthesia* **63**(7); Suppl. 1: 44S–46S.

Gosselink RA, Wagenaar RC, Rijswijk H, Sargeant AJ, Decramer ML (1995) Diaphragmatic breathing reduces efficiency of breathing in patients with chronic obstructive pulmonary disease. *American Journal of Respiratory Critical Care Medicine* **151**(4): 1136–1142.

Greenstone M (2002) Changing paradigms in the diagnosis and management of bronchiectasis. *American Journal of Respiratory Medicine* **1**(5): 339–347.

Irwin RS, Baumann MH, Bolser DC, Boulet LP, Braman SS, Brightling CE et al. (2006) Diagnosis and management of cough executive summary: ACCP evidence-based clinical practice guidelines. *Chest* **129**(1); Suppl.: 1S–23S.

Jones AP, Rowe BH (2006) Bronchopulmonary hygiene physical therapy for chronic obstructive pulmonary disease and bronchiectasis (Review). *Cochrane Library* **2** http://www.thecochranelibrary.com.

Kellett F, Redfern J, Niven RM (2005) Evaluation of nebulised hypertonic saline (7%) as an adjunct to physiotherapy in patients with stable bronchiectasis. *Respiratory Medicine* **99**(1): 27–31.

Konstan MW, Stern RC, Doershuk CF (1994) Efficacy of the flutter device for airway mucus clearance in patients with cystic fibrosis. *Journal of Pediatrics* **124**(5); Pt 1: 689–693.

Lannefors L, Wollmer P (1992) Mucus clearance with three chest physiotherapy regimes in cystic fibrosis: a comparison between postural drainage, PEP and physical exercise. *European Respiratory Journal* **5**(6): 748–753.

Lapin CD (2002) Airway physiology, autogenic drainage, and active cycle of breathing. *Respiratory Care* **47**(7): 778–785.

Macklem PT (1974) Physiology of cough. *Transactions of the American Broncho-Esophalogical Association*, pp. 150–157.

Macklem PT (1971) Airway obstruction and collateral ventilation. *Physiological Reviews* **51**(2): 368–436.

McGavin CR, Gupta SP, Lloyd EL, McHardy GJ (1977) Physical rehabilitation for the chronic bronchitic: results of a controlled trial of exercises in the home. *Thorax* **32**(3): 307–311.

Mead J, Turner JM, Macklem PT, Little JB (1967) Significance of the relationship between lung recoil and maximum expiratory flow. *Journal ofApplied Physiology* **22**(1): 95–108.

Miller DL (1963) A study of techniques for the examination of sputum in a field survey of chronic bronchitis. *American Review of Respiratory Diseases* **88**: 473–483.

Newall C, Stockley RA, Hill SL (2005) Exercise training and inspiratory muscle training in patients with bronchiectasis. *Thorax* **60**(11): 943–948.

Newhouse PA, White F, Marks JH, Homnick DN (1998) The intrapulmonary percussive ventilator and flutter device compared to standard chest physiotherapy in patients with cystic fibrosis. *Clinical Pediatrics* **37**(7): 427–432.

Patterson JE, Bradley JM, Elborn JS (2004) Airway clearance in bronchiectasis: a randomised crossover trial of active cycle of breathing techniques (incorporating postural drainage and vibration) versus test of incremental respiratory endurance. *Chronic Respiratory Disease* **1**(3): 127–130.

Patterson JE, Bradley JM, Hewitt O, Bradbury I, Elborn JS (2005) Airway clearance in bronchiectasis: a randomised crossover trial of active cycle of breathing techniques versus Acapella. *Respiration* **72**(3): 239–242.

Pryor JA (2004) Physical therapy for adults with bronchiectasis. *Clinical Pulmonary Medicine* **11**(4): 201–209.

Pryor JA (2005) *A Comparison of Five Airway Clearance Techniques in the Treatment of People with Cystic Fibrosis* PhD thesis, Imperial College London.

Pryor JA, Webber BA, Hodson ME, Batten JC (1979) Evaluation of the forced expiration technique as an adjunct to postural drainage in treatment of cystic fibrosis. *British Medical Journal* **2**(6187): 417–418.

Rose VL (1999) American Thoracic Society issues consensus statement on dyspnea. *American Family Physician* **59**(1): 3259–3260.

Rosen MJ (2006) Chronic cough due to bronchiectasis: ACCP evidence-based clinical practice guidelines. *Chest* **129**(1); Suppl.: 122S–131S.

Sackett DL, Rosenberg WMC, Gray JAM, Haynes RB, Richardson WS (1996) Evidence based medicine: what it is and what it isn't. *British Medical Journal* **312**: 71–72.

Schöni MH (1989) Autogenic drainage: a modern approach to physiotherapy in cystic fibrosis. *Journal of the Royal Society of Medicine* **82**; Suppl. 16: 32–37.

Selsby D, Jones JG (1990) Some physiological and clinical aspects of chest physiotherapy. *British Journal of Anaesthesia* **64**(5): 621–631.

Singh SJ, Morgan MD, Scott S, Walters D, Hardman AE (1992) Development of a shuttle walking test of disability in patients with chronic airways obstruction. *Thorax* **47**(12): 1019–1024.

Sutton PP, Gemmell HG, Innes N, Davidson J, Smith FW, Legge JS, Friend JA (1988) Use of nebulised saline and nebulised terbutaline as an adjunct to chest physiotherapy. *Thorax* **43**(1): 57–60.

Thompson B, Thompson HT (1968) Forced expiration exercises in asthma and their effect on FEV_1. *New Zealand Journal of Physiotherapy* **3**: 19–21.

Thompson CS, Harrison S, Ashley J, Day K, Smith DL (2002) Randomised crossover study of the flutter device and the active cycle of breathing technique in non-cystic fibrosis bronchiectasis. *Thorax* **57**: 446–448.

Tomashefski JF Jr, Bruce M, Goldberg HI, Dearborn DG. (1986) Regional distribution of macroscopic lung disease in cystic fibrosis. *American Review of Respiratory Disease* **133**(4): 535–540.

van der Schans CP (1997) Forced expiratory manoeuvres to increase transport of bronchial mucus: a mechanistic approach. *Monaldi Archives for Chest Disease* **52**(4): 367–370.

Varekojis SM, Douce FH, Flucke RL, Filbrun DA, Tice JS, McCoy KS et al. (2003) A comparison of the therapeutic effectiveness of and preference for postural drainage and percussion, intrapulmonary percussive ventilation, and high-frequency chest wall compression in hospitalized cystic fibrosis patients *Respiratory Care* **48**(1): 24–28.

Warwick WJ, Hansen LG (1991) The long-term effect of high-frequency chest compression therapy on pulmonary complications of cystic fibrosis. *Pediatric Pulmonology* **11**(3): 265–271.

Webber BA (1980) Living to the limit: exercise for the chronic breathless patient. *New Zealand Journal of Physiotherapy* **8**: 22–23.

Webber BA, Parker RA, Hofmeyr JL, Hodson ME (1985) Evaluation of self-percussion during postural drainage using the forced expiration technique. *Physiotherapy Practice* **1**: 42–45.

West JB (1997) *Pulmonary Pathophysiology* (5 edn) Baltimore: Williams and Wilkins.

Wills PJ, Wodehouse T, Corkery K, Mallon K, Wilson R, Cole PJ (1996) Short-term recombinant human DNase in bronchiectasis. Effect on clinical state and in vitro sputum transportability. *American Journal of Respiratory Critical Care Medicine* **154**(2); Pt 1: 413–417.

Wilson R (2003) Bronchiectasis. In: Gibson GJ, Geddes DM, Costabel U, Sterk PJ, Corrin B (eds) *Respiratory Medicine* (3 edn) Edinburgh: Saunders **2**: 1445–1464.

Wong JW, Keens TG, Wannamaker EM, Crozier DN, Levison H, Aspin N (1977) Effects of gravity on tracheal mucus transport rates in normal subjects and in patients with cystic fibrosis. *Pediatrics* **60**(2): 146–152.

2 Coordinated Management of a Patient in ICU with Cardiorespiratory Failure

ELIZABETH DEAN

INTRODUCTION

This ICU case exemplifies a vital role for physiotherapy and the evidence-based rationale for comprehensive patient management in the absence of mucous retention. Physiotherapy was designed to optimise long-term as well as short-term outcomes of Mrs KJ's comprehensive interdisciplinary care in the presence of the diseases of civilisation (see recent review Dean 2006a **R**; World Health Organisation 1997 **R**), and enable her return to a full life in the community, potentially at a higher level than prior to this episode of illness.

Mrs KJ is a 65 year old East Indian woman who immigrated to the UK with her husband 15 years ago. She is a retired librarian but continues to volunteer at the local library. She and her husband own an apartment in a medium-sized city in the north of England. They have family members living nearby who are highly supportive. Mrs KJ and her husband are active members of their temple, and are close to their extended family. Overall, their lifestyle is sedentary.

HISTORY OF ONSET OF PRESENT CONDITION AND OTHER RELEVANT PRE-EXISTING PATHOLOGY

On her way home from the library, Mrs KJ became short of breath on climbing the steps of her apartment building. Her husband called 999, and she was taken to the local emergency room. She reported no chest pain. Her nail bed colour was dusky. She expectorated small amounts of clear secretions tinged with bright red blood. Supplemental O_2 was administered by nasal prongs at 3 L/min. A sedative and anxiolytic agent were administered to relax her. Her temperature remained at 38 °C, respiratory rate (RR) was 35 breaths/min, heart rate (HR) 120 beats/min, and blood pressure (BP) 160/74 mm Hg. Her arterial blood gases (ABGs) were marginal and a decision was made not to intubate. Her gases deteriorated overnight however, which necessitated intubation and mechanical ventilation (assist control set at

12 breaths/min) with 60 % O_2 administered. A diagnosis of cardiorespiratory failure was made when the PaO_2 fell below 60 mm Hg and the $PaCO_2$ rose above 50 mm Hg (Shoemaker 1999 **R**).

The physiotherapist assessed Mrs KJ the morning after admission. She was resting comfortably.

Vital signs: temperature – 38 °C (slightly elevated); RR rest – 22 breaths/min; HR rest – 108 beats/min; BP rest – 155/98 mm Hg; ECG – normal sinus rhythm with occasional unifocal premature ventricular contractions (PVCs), and no apparent distress or pain other than when she was lying in one position for an extended period.

Although Mrs KJ was oriented, her arousal was reduced. She was able to purposefully and voluntarily move all limbs and change her body position with minimal assistance.

Inspection: moderately hyperinflated chest wall with reduced chest wall mobility.

Cough: moderately strong and nonproductive.

Auscultation: distant breath sounds throughout; end expiratory crackles consistent with congestive heart failure, and wheezing consistent with bronchospasm.

Heart sounds: compared with those reported on admission, heart sounds were consistent with resolving CHF; muffled heart sounds.

Extremities: cool to the touch, with evidence of healed abrasions on both legs.

Urinary output: within normal limits.

QUESTION 1

Why does risk factor assessment such as that outlined below need to be included in the physiotherapy assessment of all patients, including those in the ICU, in the contemporary health care milieu?

MAIN DIAGNOSES, AND TYPES AND EXTENT OF IMPAIRMENT AND DISABILITY

Mrs KJ's risk factors for heart disease, stroke, and diabetes were 'above average' to 'high' (Harvard University, School of Public Health 2006 **R**; Janssen et al. 2004 **A**). She was 15 kg overweight with abdominal obesity. She quit smoking six years ago. Primary diagnoses include acute respiratory dysfunction secondary to CHF and a history of coronary artery disease (CAD); severity New York Heart Association (NYHA) Classification III (New York Heart Association 2006 **R**). Co-morbidity included hypertension (Stage 2) (American Heart Association 2002 **R**), obesity (Class 1) (Expert Panel 1998 **R**) and non insulin dependent diabetes mellitus (NIDDM) (American Diabetes Association 2006 **R**). Her pulmonary function assessed two years ago was consistent with moderately severe chronic obstructive lung disease. Based on family report, Mrs KJ's aerobic capacity was low (American College of Sports Medicine 2006 **R**; McGavin et al. 1978 **A**). Normal values are based on general population norms for people living in Western countries, with no appropriate culturally-specific norms available for Mrs KJ's immigrant group.

DETAILED DESCRIPTION OF PATIENT'S PRESENTING SYMPTOMS AND PROBLEMS

Physiotherapy Diagnosis: impaired O_2 transport with incapacity to maintain adequate blood gases and gas exchange without ventilatory support and supplemental O_2. Her capacity to engage in her activities of daily living (ADLs) and socially participate has been severely compromised over the past two years, and particularly the past several weeks.

Analysis of O_2 demands (during this episode of ICU care):

- Breathlessness on minimal exertion (but remains on assist control mechanical ventilation; initiating all breaths herself compared with when she was first mechanically ventilated).
- Myocardial demands to meet increased systemic O_2 demands.
- Respiratory demands increased, due to increased RR to help increase gas exchange.
- Metabolic demands increased due to slighted elevated temperature.
- Anxiety (but remains controlled with medication).
- Demands to meet needs of increased body weight.
- Demands of healing and recovery.

QUESTION 2

The World Health Organisation's International Classification of Function includes additional levels of assessment for physiotherapists, including in the ICU. How does this affect your assessment and management of Mrs KJ?

RESULT OF PHYSICAL EXAMINATION AND LABORATORY TESTS AND INVESTIGATIONS

The findings are classified according to the International Classification of Function (World Health Organisation, 2002 **R**), namely, social participation (handicap), activities (disability), and structure and function (impairment).

Social participation (history from husband and family)

Mrs KJ has a supportive husband and family. A health-related quality of life questionnaire modified for use in acute settings, Short Form-36, (Ambrosino 2002 **A**; Short Form-36 2000 **R**) was completed by proxy (Hoffhuis et al. 2003 **A**), that is, by her husband, with Mrs KJ's consent. The initial score reflected her functional capacity prior to this illness episode. The questionnaire has also been adapted for use with people of East Indian descent to maximise its validity in this population. Her scores (23/50 on the physical health scale and 32/50 on the mental health scale) serve as an outcome measure, hence, a guide for Mrs KJ's eventual return to her family and community life, and to an improved quality of life. Although she has reduced her social activities over the past year, Mrs KJ has maintained her activities and responsibilities as wife and homemaker, and continues to serve as a volunteer librarian three mornings a week. She enjoys having her three grandchildren over to her home, but has been finding it exhausting over the past two months. She is a regular visitor to a friend with a disability. She did serve as a volunteer in the office at her temple but discontinued last year because she felt it was too much.

Activities (composite activities based on history with family and analysis of activities prior to this episode of illness)

For health reasons, Mrs KJ's social and recreational activities have become progressively restricted, for example, she is less able to get back and forth to visit her daughter and family, and to get to the temple and the library.

Structure and function

Blood work: on an F_IO_2 of 50 % (reduced with progressive improvement in ABGs since admission) while mechanically ventilated – PaO_2 85 mm Hg, $PaCO_2$ 47 mm Hg, pH 7.42, HCO_3 30 mEq/L and SaO_2 94 %.

X-rays and scans: chest X-ray – chest wall hyperinflation. The classic butterfly sign of CHF was present, and this pattern has shown progressive clearing with the administration of diuretic therapy since admission. Microatelectasis is apparent centrally and in the bases.

QUESTION 3

Why does the physiotherapist need to consider oxygen transport as a whole, rather than focusing on airway clearance?

The steps in the O_2 transport pathway include the airways and lungs, the pulmonary circulation, the blood, the heart and its electromechanical coupling, the peripheral circulation, and O_2 extraction at the tissue and muscle levels. O_2 transport is a function of Mrs KJ's capacity to meet her O_2 demand given her capacity to supply O_2. One or more steps in the O_2 transport pathway can be impaired and/or threatened by four primary factors (Dean 1994 **R**): recumbency and restricted mobility; extrinsic factors related to her care (for example, side effects of pharmacologic agents); intrinsic factors related to the patient (for example, obesity and sedentary lifestyle); and the patient's underlying pathophysiology. Initially the patient is medically stabilised, which supports healing and repair, and regaining of homeostasis with optimal rest and sleep periods. Inotropic support, sedation to reduce arousal and undue metabolic demand, and diuretics were administered.

PHYSIOTHERAPY GOALS

Mrs KJ is limited by arthritic pain and deconditioning, combined with the effects of atherosclerosis, ischemic heart disease rendering her heart hypoeffective as a pump, and emphysematous lung changes secondary to COPD impairing respiratory mechanics and gas exchange. This latter acute episode, resulting from CHF, has worsened her gas exchange to the point of needing supplemental O_2 and ventilatory support. The physiotherapist needs to ensure recumbency and bed rest are minimised to limit further aerobic compromise, deconditioning and complications (Allen et al. 1999 **R**; Bolton 2001 **R**; Saltin et al. 1998 **A**). This is judiciously balanced with Mrs KJ's requirements for rest. The short-term goals need to address Mrs KJ's life-threatening priorities related to O_2 transport and prevention of complications (Dantzker 1991 **R**; Dantzker et al. 1991 **R**). However, this is the first component of the continuum of physiotherapy, with a view to Mrs KJ's achieving a higher level of health than prior to this episode of illness, in the months to come.

Short-term goals and strategies

- Prevent O_2 transport deficits and systemic complications including neuropathies, myopathies, and skin breakdown due to recumbency, restricted mobility and reduced psychosocial wellbeing.
- Optimise O_2 transport (from airways and lungs to the tissue level, including optimising oxidative enzymes at the muscle tissue level to augment O_2 transport secondary to improved O_2 extraction at this level).
- Minimise undue work of breathing and work of the heart.
- Stabilise hemodynamic status.
- Optimise cardiac output.
- Optimise sympathetic nervous system activity, with a view to helping reduce inotropic medication.
- Commensurate with patient's level of understanding and readiness, reinforce positive health choices and behaviours including living in a smoke-free environment, optimising nutrition, weight control, physical activity, a modified exercise programme, and sleep and stress management.
- Reduce anxiety and promote physical comfort (generally and specifically related to being mechanically ventilated); enable Mrs KJ to communicate when ventilated.
- Identify readiness to wean with the team, and participate in the weaning process and post-weaning period to maximise weaning success and minimise risk of re-intubation.
- Work toward replacing invasive with noninvasive mechanical ventilation to minimise risk of failure to wean, a risk in people with COPD.
- Involve the family from the outset to optimise psychosocial support and recovery rate (Jones et al. 1994 **R**).

Intensive care unit (days one to five)

Days One and Two. Although her arousal is reduced, Mrs KJ is alert and oriented. She has been medicated to help reduce myocardial work and the work of breathing, and metabolism overall (Weissman et al. 1984 **A**; Weissman et al. 1989 **A**; Weissman & Kemper 1993 **A**; Weissman et al. 1994 **A**). The nursing staff has instituted a two-hourly turning regimen to help reduce multisystem complications associated with bed

Table 2.1. Hemodynamic effects of specific body positions

Position	Effects
Upright	Caudal displacement of fluid shifts in the body (Blomqvist & Stone 1963 R; Gauer & Thron 1965 R; Sandler 1986 R).
	Compensatory increase in heart rate.
	↓ Myocardial work (Langou et al. 1977 A; Levine & Lown 1952 A).
	↑ Peripheral vascular resistance.
	↑ Threshold for anginal pain (Prakash et al. 1973 R).
	Compression of the viscera on the dependent hemidiaphragm (Lange et al. 1988 A).
Side lying	↑ End diastolic ventricular pressure on the dependent side (Lange et al. 1988 A).
	Optimal ventilation to perfusion matching in the upper one-third of each lung in side lying (Kaneko et al. 1966 A).
	PaO_2 greater in side lying than supine (Clauss et al. 1968 A).
	Arterial blood gases improved in patients with unilateral lung disease with unaffected lung down (Remolina 1981 A; Sonneblick et al. 1983 A).
	Enhanced mucociliary transport.
	Cardiac compression and reduced compliance of adjacent lung field (Lange et al. 1988 A).
Left	↑ Cardiac compression.
Right	Potentially less cardiac compression.
Prone	Potentially ↑ cardiac compression and improved gas exchange (Chatte et al. 1997 A).
Supine	Cephalic displacement of fluid shifts in the body (Blomqvist & Stone 1963 A).
	↑ Preload and afterload of the right side of the heart.
	↓ Left ventricular volume and preload (Prefaut & Engel 1981 A).
	Cephalic displacement of the abdominal viscera (Barach & Beck 1954 A).
	Pulmonary arteriovenous shunt (Ray et al. 1974 A).

Note: for further references see reviews Dean 2006c R; Doering 1993 R.

rest, including decreased insulin sensitivity (Mikines et al. 1991 A), compounding her existing problem, and prevent critical illness polyneuropathy and myopathy (Bolton 2001 R; Heaton 1999 R; Kollef 1999 A). The ICU team's plan is to observe her progress and hemodynamic stability for 48 hours, and then consider weaning from mechanical ventilation. Although mobilisation is not indicated at this time, the physiotherapist is regularly assessing the patient to determine when a window of opportunity arises for body positioning to be instituted to address her O_2 transport deficits. Tables 2.1 and 2.2 show the pulmonary and hemodynamic effects of some common body positions. Understanding the hemodynamic consequences as well as the pulmonary effects of different body positions, including perturbation of the distribution of ventilation (Jones & Dean 2004 A; Kim et al. 2002 A), is critical to using body positioning discriminately for its beneficial effects, and understanding adverse effects. Initially, the goal is to get this patient 'upright and moving', given

Table 2.2. Pulmonary effects of specific body positions

Position	Effects
Upright	↑ Lung volumes and flow rates (Svanberg 1957 A). Optimal length tension ratio of the respiratory muscles (Druz & Sharp 1991A; Sharp et al. 1980 A).
Side lying	Anteroposterior excursion accentuated at the expense of laterocostal expansion of the dependent side. Alveolar volume favoured to the non dependent lung. Ventilation and perfusion favoured in the dependent lung. Functional residual capacity midway between sitting and supine. In bilateral lung disease, arterial blood gases worse in right side lying and better in left side lying (Zack et al. 1974 A). Improved arterial blood gases, tidal volume and lung compliance (Douglas et al. 1977 A; Gillespie & Rehder 1987A; Ibanez et al. 1981 A; Langer et al. 1988 A; Wagaman et al. 1979 A).
Prone	Prone abdomen free superior to prone abdomen restricted (in neonates) (Mellins 1974 A). Improves ventilatory function (V/Q) and efficiency in patient with lung injury (Shickinohe et al. 1991 A).
Supine	Visceral compression of the hemidiaphragms, reduces lung volumes (Svanberg 1957 A). ↓ Functional residual capacity, increase airway closure (Hsu & Hickey 1976 A; Sjostrand 1951 A). Lung volume effects are accentuated with ageing (Langer et al. 1988 A). ↑ Closing volume of the dependent airways (Leblanc et al. 1970 A). ↑ Airway resistance. ↓ Lung compliance (Sasaki et al. 1977 A). ↓ Intrathoracic volume and ↑ intrathoracic pressure. Chest wall compression in the anteroposterior plane and limited chest wall excursion (Behrakis et al. 1983 A; Craig et al. 1971 A; Don et al. 1971 A). Altered respiratory muscle function (Roussos et al. 1976 A).
Head down and forward leaning	In patients with flattened diaphragms, head down can augment diaphragmatic function and reduce shortness of breath (Barach & Beck 1954 A; De Troyer 1983 A).

Note: for further references see review in Dean 2006c R.

that recumbency will contribute to reduced blood volume and potential for thrombus formation (Convertino 1992 A).

QUESTION 4

Body position and mobilisation are powerful tools to counter bed rest deconditioning as well as address oxygen transport deficits. What factors determine the use of body positioning vs. mobilisation, as well as their joint use in any given treatment for Mrs KJ?

Based on this literature and Mrs KJ's assessment, a decision can be made regarding the optimal body position in terms of reducing undue metabolic demands by improving gas exchange (Dean 1985 **R**; Dean & Ross 1992 **A**). A given body position can be maintained for as long as gas exchange is being optimised, but this does not usually exceed two hours. Timing of intervention is crucial. Changing body position and maintaining a given body position can have positive or negative effects, so the patient must be closely monitored (Dean 2006c **R**). Extreme body position changes, if well tolerated, are preferable in that they better simulate the normal gravitational stressors on the cardiopulmonary unit (Piehl & Brown 1976 **A**). This information can also be used as a basis for clinical decision making in a subsequent treatment.

At ward rounds, the physiotherapist raises two concerns: one, the opiate being used to reduce Mrs KJ's O_2 consumption is reducing her arousal and capacity to cooperate with assessment and potential intervention. Two, Mrs KJ's body position is not ideal for her treatment. The physiotherapist proposes to gravitationally challenge Mrs KJ first with legs non dependent and then progress to dependent. The team supports the decision to try another medication that is associated with less grogginess, and to observe how she responds hemodynamically, specifically in terms of HR, BP, cardiac output, and ECG, to being positioned upright. The physiotherapist also queries whether noninvasive ventilation, for example, nasal ventilation, may be of more benefit (Bott et al. 1993 **A**; Kramer et al. 1995 **A**; Ram et al. 2003 **R**) and more cost effective (Plant et al. 2003 **A**).

Day Three. Mrs KJ's chest X-ray shows bilateral basal atelectasis; L side > R side. There is no evidence of mucous retention. Her urinary output remains acceptable.

QUESTION 5

What parameters and factors does the physiotherapist need to consider to guide the prescription of mobilisation for Mrs KJ?

Days Four and Five. Mrs KJ is instructed in general relaxation, relaxed breathing, and supported coughing (huffing to minimise increasing intrathoracic pressure), to reduce undue energy expenditure during progressive mobilisation with ongoing monitoring (Dean & Ross 1992 **A**). Conventional so-called diaphragmatic breathing has been questioned in that it has been associated with reduced breathing efficiency in patients with COPD (Gosselink et al. 1995 **A**). Table 2.3 shows evidence for mobilisation as the single most important ICU intervention, and its major benefits on priming and conditioning O_2 transport. Detailed monitoring and the basis for progressing mobilisation for patients in the ICU with primary cardiopulmonary dysfunction have been reviewed in detail elsewhere (Dean & Perme 2006 **R**; Holten 1972 **A**; Wenger 1982 **R**; Wong 2000**A**; Yohannes & Connelly 2003 **A**). Much like aerobic conditioning for people in health, such conditioning is needed in those with threats and deficits to O_2 transport. However, the mobilisation or exercise parameters (type of mobilisation, intensity, duration, frequency and course) need to be modified (American College of

Table 2.3. Acute effects of mobilisation (Dean 2006b R)

Pulmonary System
↑ Regional ventilation
↑ Regional perfusion
↑ Regional diffusion
↑ Zone 2 (for example, area of ventilation perfusion matching)
↑ Tidal volume
Alters breathing frequency
↑ Minute ventilation
↑ Efficiency of respiratory mechanics
↓ Airflow resistance
↑ Flow rates
↑ Strength and quality of a cough
↑ Mucociliary transport and airway clearance
↑ Distribution and function of pulmonary immune factors

Cardiovascular System
Hemodynamic effects:
   ↑ Venous return
   ↑ Stroke volume
   ↑ Heart rate
   ↑ Myocardial contractility
   ↑ Stroke volume, heart rate and cardiac output
   ↑ Coronary perfusion
Hematologic effects:
  Stimulates ↑ in circulating blood volume
  Stimulates reduced coagulation and platelet aggregability
  Peripheral circulatory effects
  ↓ Peripheral vascular resistance
  ↑ Peripheral blood flow
  ↑ Peripheral tissue oxygen extraction
  ↑ Circulatory transit times
  ↓ Circulatory stasis

Lymphatic System
↑ Pulmonary lymphatic flow
↑ Pulmonary lymphatic drainage

Neurological System
↑ Drive to breathe
↑ Arousal
↑ Cerebral electrical activity
↑ Stimulus to breathe
↑ Sympathetic stimulation
Primes postural control and reflexes

Neuromuscular System
↑ Regional blood flow
↑ Oxygen extraction

Musculoskeletal System
Stimulates osteogenesis
Strengthens connective tissue

Endocrine System
↑ Release, distribution, and degradation of catecholamines
Stimulates endorphin production

(Continued)

Table 2.3. Acute effects of mobilisation (Dean 2006b R) (*Continued*)

Genitourinary System
↑ Glomerular filtration
↑ Urinary output (Foley catheter drainage)
↓ Renal stasis

Gastrointestinal System
↑ Gut motility
↓ Gastrointestinal transit time
↓ Constipation

Integumentary System
↑ Cutaneous circulation for thermoregulation

Sports Medicine 2006 **R**; Dean 2006b **R**). Equipment such as a rollator is used to maximise ventilation, gas exchange and aerobic capacity (Probst et al. 2003 **A**).

Type of mobilisation – progressive exercise, including sitting up in bed, legs dependent; standing and shifting weight; transfer to chair; chair exercises; and walking with the ventilator.

Intensity – HR below HRrest plus 20 beats/min.

Duration – interval mobilisation or exercise protocols are used to avoid inappropriate exercise responses, and promote rest and recovery during the mobilisation period.

Frequency – as often as can be tolerated safely; the more acutely ill the patient, the less intense and shorter the sessions, but the greater the frequency.

Course – progressive mobilisation continues until the patient is discharged from the ICU. Her care is seamlessly assumed by a physiotherapist on the ward, and then in the community.

In the earlier phases of ICU care, body positioning is used to simulate the normal physiologic body position of being 'upright and moving'. However, to initiate mobilisation the patient needs to be relatively hemodynamically stable, both for safety and to ensure that she has the hemodynamic reserve capacity to respond to the exercise load. Thus, mobilisation is initiated slowly and progressively with close hemodynamic monitoring. Both mobilisation and body position changes are performed gradually to ensure pre-set criteria related to her hemodynamic stability are not exceeded.

Mrs KJ tolerates sitting up (erect) with feet over the bed (supported), with one person assisting, for 10 minutes in the morning and 20 minutes in the afternoon. Her HR and BP remain within acceptable levels (10 to 15 % of baseline, returning quickly to resting levels within a few minutes of cessation of movement). She is progressed to standing in the evening, with weight shifting from one foot to the other, and sits in the bedside chair for one hour. This activity is progressed slowly and with no breath holding or heavy gasping. Similarly, HR, BP and RR remain within 15 % of baseline levels. No dysrhythmias, including PVCs, are observed. Chair exercises are conducted at the beginning and at the end of this time for 15 and 20 minutes respectively. Smooth coordinated movements are encouraged, with vital signs remaining within safe and therapeutic levels. She is able to perform: arm elevations (full range of motion), three sets of 10 repetitions; left and right side flexion (15 times); and trunk rotation (15 times to each side). This programme of moderate intensity is performed with coordinated deep breathing and coughing manoeuvres. Breathing control facilitates venous return and cardiac output, primes sympathetic nervous stimulation of the peripheral blood vessels, and stimulates surfactant production and distribution secondary to stretching the lung parenchyma. She is encouraged however to avoid exhaling below end-resting tidal volume to minimise closure of the dependent airways. Coughing (huffing with glottis open to minimise excessive hemodynamic response) is encouraged every five minutes or as required. Coughing requires a large inspiratory volume followed by increased flow rates. Due to the monotonous pattern of tidal ventilation on mechanical ventilation, 'more normal' mucociliary transport is thereby facilitated.

Days Six and Seven. Mrs KJ is transferred to a bedside chair several times during the day under the supervision of the physiotherapist. With each attempt, the reduced level of physical support is recorded as this is an important outcome of physiotherapy. She is continuously monitored throughout treatments to ensure the interventions are both safe and therapeutic. No more than six PVCs/min. are acceptable during treatment (Dubin 2000 **R**) and the intensity of treatment is titrated to her self-reported tolerance, and to maintain HR with 20 bpm and BP within 20 to 30 mm Hg of systolic BP. On transferring, she stands erect and shifts her weight from side to side for three minutes with increasingly less support from the physiotherapist. She sits in the bedside chair for one hour in the morning and for two hours in the afternoon. Chair exercises, coordinated with breathing control, include forward flexion and extension, left and right lateral bending, and left and right trunk rotation. Lower extremity exercises include alternate lifting left and right knees, and left and right control knee flexion and extension. Upper extremity exercises include shoulder flexion and extension, and abduction and adduction. Note: erect body postures are encouraged to maximise her pulmonary function and respiratory muscle contraction.

QUESTION 6

What is physiotherapy's role in weaning a patient from mechanical ventilation and what monitoring needs to be incorporated to ensure this is performed safely and at the right time?

Day Eight. Mobilisation, including walking such as that prescribed for Mrs KJ, has long been proposed as a means of facilitating weaning from mechanical ventilation (Burns & Jones 1975 **A**). With a progressive mobilisation programme prescribed within safe and therapeutic limits, Mrs KJ's aerobic capacity and gas exchange are showing signs of being more efficient. Her ABGs have remained within acceptable limits and stable for 72 hours. Mechanical ventilation is limited as much as possible for all patients and particularly for those with COPD given their abnormal drive to breathe. However, because respiratory muscle fatigue is a cause of respiratory failure in people with COPD (Macklem & Roussos 1977 **A**), such fatigue needs to be ruled out in Mrs KJ. Established evidence-based guidelines for extubation are implemented to maximise its success (MacIntyre et al. 2001 **R**). Morning vs later extubations with the patient alert and upright may be associated with improved outcomes including reduced risk of re-intubation. The physiotherapist participates in the weaning and ensures that Mrs KJ does not desaturate during the procedure. Breathing control and huffing are encouraged immediately post extubation and every two to three hours, coordinated with her mobilisation programme. She remains on O_2 by mask for the remainder of the day and then this is replaced with nasal prongs. The physiotherapist follows her closely throughout the day to ensure her arterial saturation (assessed with pulse oximetry) and blood gases remain at acceptable levels, and that her vital signs and breathing rate are also within acceptable levels. She shows no signs of unusual breathlessness, chest discomfort or other distress. Optimal resting body positions for people with stable COPD that augment the respiratory mechanics efficiency, have been proposed to be more physiologic than attempting to reduce the work of breathing with breathing exercises (Jones et al. 2003 **A**). Thus, as Mrs KJ's condition becomes less acute, body positioning in conjunction with increasing her mobilisation level is exploited to improve breathing efficiency and sustained reduced work of breathing.

Her transfer to chair and standing weight-transferring exercises are well tolerated, and monitored closely to ensure there is no deterioration. She tolerates two hours in the bed-side chair in the morning and three hours in the afternoon. She completes her exercise programme with no signs of unusual distress or desaturation. Mrs KJ is receiving O_2 by nasal prongs at 2 L/min. Exercise termination criteria include desaturation to 90 %, HR increase more than 20 beats greater than resting HR, or BP increase greater than 20 mm Hg, or any abnormal change in ECG. If any one of these occurs, Mrs KJ rests. With increasing levels of exercise stress, caution continues to be observed (monitoring and supplemental O_2 adjustment) given the inconsistent findings on exercise-induced desaturation in patients with severe COPD and ECG changes (Jones et al. 2006 **A**).

QUESTION 7

What is the justification for the physiotherapist including lifestyle recommendations and follow-up in this ICU case?

Day Nine. A component of Mrs KJ's comprehensive programme is risk factor assessment so that risk factor modification interventions can be prescribed with a view to reducing each of her modifiable risk factors. Internationally recognised standards for cardiac rehabilitation Phase I include education about lifestyle (nutrition, weight control, physical activity and structured exercise), energy conservation, sleep and stress management, and medications. Smoking cessation recommendations are also a component of health education in Phase I. Although Mrs KJ has not smoked for many years, she is in contact with second hand smoke in the family, and this warrants being addressed.

In Phase I, the patient is progressed through incremental levels of physical activity and exercise with increasing metabolic demand (Cardiac Rehabilitation 1998 **R**). The discharge goal is to have the patient safely walk up and down one flight of stairs with vital signs and perceived exertion within acceptable levels for that individual. The indications and side effects of Mrs KJ's medications have been discussed with herself and her family, and have also been written down. In addition, means of ensuring adherence with their administration are discussed with Mrs KJ by various members of the team. The physiotherapist in the community will work closely with the GP and community nurse, to ensure medication is reduced as indicated, commensurate with Mrs KJ's weight loss and improved physical work capacity. She will be followed closely to ensure that the transition to home and community is seamless.

Air quality – Mrs KJ has compromised ventilatory reserve, thus minimising ventilatory stress is a priority. She lives in an urban area, so is exposed to poor air quality. Her son-in-law smokes but has been considering quitting.

Nutrition – the nutritionist will conduct a seven-day eating record; Mrs KJ's eating patterns prior to this episode of care are recorded to establish a baseline.

Weight control (self-monitoring) – the nutritionist and physiotherapist will monitor. The nutritionist has designed a balanced, nutritious weight loss programme that considers Mrs KJ's ethnic preferences. The physiotherapist discusses with the nutritionist and Mrs KJ the metabolic demands of each day's physical activity and exercise programme, in preparation for her discharge.

Physical activity and exercise – these are progressed. Distance walked three times daily, including number and duration of rests, is recorded. Strength training includes 1 lb weights in each hand, and 2 lb weights attached to each ankle, for her chair exercise programme; five repetitions of three of each exercise for each upper and lower extremity muscle group.

Stress management and sleep quality – patients in ICUs have poor quality sleep (Peruzzi 2005 **R**; Walder et al. 2000 **A**), thus the team coordinates each member's time with Mrs KJ to promote optimal sleep, minimising sleep disruptions, particularly through the night.

Day 10. Mrs KJ is transferred to the general ward for reassessment and discharge planning with the interdisciplinary team. Pre-discharge risk factors for the diseases of civilisation, including ischemic heart disease (risk of another event), stroke, diabetes and cancer are assessed based on established questionnaires, and an education plan is developed with the team, including the physiotherapist.

Day 12. The physiotherapist and other team members, including the social worker and occupational therapist, meet with Mrs KJ and close family members to discuss Mrs KJ's discharge plan. Her home has been adapted and was viewed as safe by a public health occupational therapist this past year.

The six minute walk (SMW) test is administered (McGavin et al. 1978 **A**; Noonan & Dean 2000 **R**).

Day 13. The SMW test is repeated to ensure that the results were valid and reliable.

QUESTION 8

What cultural factors need to be considered to modify lifestyle recommendations?
Give the rationale.

Day 14. The physiotherapist completes the final discharge assessment for the community team and outlines the follow-up that needs to be instituted to ensure that Mrs KJ's short-term goals are sustained, and that the long-term goals are being instituted with a view to maximising her functional capacity and social participation, minimising her health risk factors (Hu et al. 2004 **A**) and the need for medical or potentially surgical intervention, and minimising her medication.

Mrs KJ's native culture needs to be considered as an important factor in her care with respect to her health beliefs and behaviours, and beliefs about her condition and self-efficacy regarding its life-long management. Indian culture is distinct from Western culture in that it is collectivistic vs individualistic, tends to respect people in positions of authority, and is considered high vs low context (Hofstede 1980 **R**; Singelis et al. 1995 **R**). In a practical sense, her orientation and goals relate to her family rather than her personal interests. She is eager to resume her responsibilities as wife and grandmother, and activities associated with community service in general (through library work and her work at the temple). She respects the knowledge of her interdisciplinary health care team, and their interest in designing a life-long health programme for her. She is receptive to their recommendations. She is also interested in the traditional health care practices of Ayurvedic medicine practised in India and would like to integrate them into her programme. Yoga and meditation may have health benefits (Oken et al. 2006 **A**), and some benefits specifically related to the control of hypertension and blood glucose when coupled with exercise.

In terms of health education, explicit information may be more effective than generalisations given India is considered a high context culture compared with the West. The degree to which Mrs KJ's world view reflects that of a high context culture needs to be established. She is cautioned about using traditional herbal remedies at this time given that her management during this episode of care has been Western. Should traditional remedies have interest for her, she should discuss this with both her GP and traditional practitioner to avoid confounding the effects of two medical approaches and risking potential adverse interactions and side effects.

Mrs KJ's learning needs are assessed. Although she is proficient in speaking, reading and writing English, her culture needs to be considered in the design of the health programme if she is to adhere long-term and derive life-long benefit. With respect to her learning style, Mrs KJ prefers to write things down in her own words, so that they make sense to her. She wants to be involved with developing record sheets for her medications, her nutritional plan and weight loss regimen, and her physical activity and exercise programmes. These are formatted in a way consistent with what appears logical and convenient to her. She is pleased to adhere to the programme, and to report back to the physiotherapist and other team members in the next two weeks. She is highly responsive to the idea of reporting back to the physiotherapist, and having an opportunity to consult with a health professional if any untoward changes occur. The physiotherapist reinforces medication teaching by the discharging nurse to ensure

Mrs KJ takes the prescribed medications and understands the potential consequences of not doing so. In addition, Mrs KJ has learned how to use the glucometer that her husband has purchased on the recommendation of the team, and has been keeping a log book of her diet, physical activity, and blood sugar levels in the three days prior to discharge. She is to maintain the log book and present it at her physiotherapy follow-up visits, and to her physician and community nurse.

One of the most important components of Mrs KJ's discharge plan will be the progressive exercise programme that was initiated on day two of her ICU stay. Now she is stable, the parameters for her flexibility, aerobic and strength programmes are prescribed based on assessment-based needs at discharge, and are progressed based on re-testing. Consideration was given to including inspiratory muscle training (Scherer et al. 2000 A). Given Mrs KJ is medically stable and interested in lifestyle modification, a decision was made to monitor inspiratory muscle strength and use this as one of the outcome measures over the next three months.

Her exercise plan is designed to exploit the well-established long-term multisystem benefits (see Table 2.4). Achieving these benefits requires progressive training, through which she will develop both central and peripheral, and metabolic adaptations (Braith & Vincent 1999 A; Expert Panel 1998 R; Hoppeler & Fluck 2003 A). Consideration is also given to the sustainability of her programme (Lennon et al. 2004 A).

Long-term and preventive goals and strategies:
- Optimal health through optimal diet (nutrition and weight control) (Ornish 1998 A; Ornish et al. 1998 A) and physical activity (Sato 2000 R).
- ↓ Cardiac symptoms.
- ↓ Shortness of breath.
- Secondary prevention of heart disease, acute exacerbation of pulmonary dysfunction, hypertension, and type 2 diabetes.
- Promotion of life-long health behaviours (with follow-up and reassessment in four weeks) including:
  − Smoke-free environment and heart and lung health.
  − Optimal nutrition.
  − Optimal weight control.
  − Regular physical activity.
  − Prescribed exercise programmes:
    - Flexibility − body positioning monitoring − erect standing position taught, and optimal biomechanics during sitting and lying. Mrs KJ performs several selected yoga exercises for 15 minutes in the morning and evening. The exercises were specifically chosen by the physiotherapist, with Mrs KJ's agreement, to focus on upper extremity, chest wall and spinal flexibility. She was cautioned to repeat these exercises slowly several times without straining. She was instructed in breathing control and how to coordinate breathing with each exercise.
    - Aerobic training − physical activity − Mrs KJ was instructed in ways to progressively increase her daily activity with the use of a pedometer. From a baseline established on her last day in hospital, the physiotherapist instructed her to begin with 700 steps a day, and progress 100 steps each week until the follow-up

Table 2.4. Long-term or chronic effects of exercise (Dean 2006b R)

Cardiopulmonary System
↓ Submaximal minute ventilation
↑ Respiratory muscle strength and endurance
↑ Collateral ventilation
↑ Pulmonary vascularisation
↓ Rating of perceived exertion or breathlessness at submaximal work rates

Cardiovascular System
↑ Myocardial muscle mass
↑ Myocardial efficiency
Exercise-induced bradycardia
↑ Stroke volume at rest and submaximal work rates
↓ Resting heart rate and blood pressure
↓ Submaximal heart rate, blood pressure and rate pressure product
↓ Submaximal perceived exertion and breathlessness
↑ Efficiency of thermoregulation
↓ Orthostatic intolerance when performed in the upright position

Hematologic System
↑ Circulating blood volume
↑ Number of red blood cells
Optimises hematocrit
Optimises cholesterol
↓ Blood lipids

Central Nervous System
↑ Sense of well-being
↑ Concentration

Neuromuscular System
Enhance neuromotor control
↑ Efficiency of postural reflexes associated with type of exercise
↑ Efficiency of reflex control
↑ Movement efficiency and economy

Musculoskeletal System
↑ Muscle vascularisation
↑ Myoglobin
↑ Muscle metabolic enzymes
↑ Glycogen storage capacity
↑ Biomechanical efficiency
↑ Movement economy
Muscle hypertrophy
↑ Muscle strength and endurance
↑ Ligament tensile strength
Maintains bone density

Endocrine System
↑ Efficiency of hormone production and degradation to support exercise
  ↑ Insulin sensitivity

Immunological System
↑ Resistance to infection

Integumentary System
↑ Efficiency of skin as a heat exchanger
↑ Sweating efficiency

reassessment. Intensity: 3–5 on the Borg scale (0, or no breathlessness at all, to 10, or maximal breathlessness) (Borg 1998 **R**). Over 9,999 steps a day is consistent with an active lifestyle and health benefits (Tudor-Locke & Bassett 2004 **A**). Between 7,500 and 9,999 is consistent with a somewhat active lifestyle, which provides a goal for Mrs KJ.

- Aerobic training – prescribed exercise programme – Mrs KJ was instructed in a walk-rest programme: beginning with five to ten minutes of walking followed by two minutes rest (3 cycles) in the morning and afternoon for two weeks (initially with her husband), and then progressing to 15–20 minutes over the following two weeks. Intensity: 3–5 on the Borg scale (0, or no breathlessness at all, to 10, or maximal breathlessness) (Borg 1998 **R**). She will be then reassessed.
- Strength training – weeks one and two: morning and afternoon; three sets of three repetitions of 1 lb weights in each hand for controlled shoulder flexion and extension, and abduction and adduction, and of 2–3 lb weights on each ankle for controlled knee extension and flexion. Breathing control and no straining. Breathlessness scale should remain below 2–3; weeks three and four: progress repetitions to three sets of five repetitions.
  – Optimal sleep – optimise quality and quantity of her night's sleep. Avoid tea or coffee or other caffeinated beverages in the evening. Recommend engaging in quiet activities after 8 pm. Develop a bedtime routine.
  – Stress management – Mrs KJ will enroll in a weekly yoga (beginners) and meditation class with her husband, and practise every day.
- Optimise health of her husband and potentially extended family, as well as herself.
- Follow-up plan: Mrs KJ to be followed by community physiotherapist, who will advise her on appropriate community resources available to her, including those with East Indian clients and culturally-appropriate programmes.
- Review of home accessibility and safety including access, rugs and carpets, bathroom accessibility (including toilet and bath access), stairs, cupboard and storage organisation, and access to the items she needs.
- Home help to be arranged for the short-term until both Mrs KJ and her husband are able to assume their home management responsibilities.
- Review of home and community accessibility, and capacity to be mobile in her community (for example, visit her daughter and family, go to the shops, to the library, and access her own home).
- Arrange for periodic follow-up (with the first follow-up in one month) and provision of contact number if she runs into difficulty between physiotherapy visits.
- Reduce risk factor for acute episodes of heart disease and lung disease (risk category rated as high for both), and reduce risk categories for stroke, diabetes, and osteoporosis.
- Minimise the need for invasive intervention including visits to her doctor.
- Minimise the need for medications (work with her GP so medications can be minimised as much as possible as Mrs KJ demonstrates specific health benefits from her life-long health programme, for example, normalised BP and blood sugar, and reduced work of breathing).

QUESTION 9

Consider the stages of behaviour change below. How would you rate Mrs KJ's stage and what factors would you consider to shift her to a higher level of readiness?

Mrs KJ's activity levels, and her readiness to change with respect to her nutrition and weight loss, are assessed (Prochaska & DiClemente 1982 **R**). The stages of change include:

- Pre-contemplative (not ready to change at this time).
- Contemplative (thinking about changing one or more health behaviours).
- Preparation (preparing to institute a change by one or more identifiable actions).
- Action (actively engaging in the health behaviour change).
- Maintenance (health behaviour changes have been well established, and have become a way of life).
- Mrs KJ is at the Preparation stage in terms of readiness to change in the primary health categories: air quality, nutrition, weight loss, physical activity, structured exercise, sleep and stress reduction. She is motivated by her role in the family, and being able to contribute to her community.

CONCLUSION

This case illustrates an integrated evidence-based physiotherapy management approach, in conjunction with team members, aimed at preventing and resolving Mrs KJ's life-threatening O_2 transport risks and deficits. The case then exemplifies integrated physiotherapy care along the continuum from acute medically-unstable to chronic medically-stable and the return of Mrs KJ to the community with an optimal quality of life. From the outset, the physiotherapist considers Mrs KJ's needs at home and in the community, and the requirements for her eventual return to optimal social participation. Mrs KJ has serious life-threatening conditions (heart and lung disease combined, and hypertension) in addition to obesity and glucose intolerance. With integrated physiotherapy management and early discharge, a life-long health plan can be designed in conjunction with her interests and needs. Such a plan increases the probability of Mrs KJ achieving and sustaining an optimal level of health, and preventing or delaying further episodes of serious illness, which have the potential for being less severe and with faster recovery. Being committed to the exploitation of noninvasive care to the highest degree possible, the physiotherapist aims to reduce the need for invasive care as much as possible, or at least reduce Mrs KJ's need for medication and invasive procedures in the short- and long-term. Reducing her need to visit her doctor, be admitted to hospital, or for medication are important physiotherapy outcomes. The GP and physiotherapist need to work together to ensure that noninvasive care is being exploited maximally in the interest of the patient's short- and long-term health. Medications, for example, that impact Mrs KJ's functional capacity need to be appropriate and optimally beneficial. If she adheres to the medication regimen, the medications should have maximal benefit with minimal side effects or risks. The noninvasive practices of physiotherapy warrant being exploited in the ICU, which is high-tech, highly invasive, and costly. Particularly in this setting, invasive care and noninvasive care need to complement each other, to minimise unnecessarily invasive care (procedures and medications) and its risks. This is achieved with coordinated team work and respect for the contribution of each member of the ICU team. The team needs to consider the quality of a patient's life after the ICU episode from the outset.

Finally, although Mrs KJ's health programme may appear ambitious, small improvements in Mrs KJ's physiologic capacity can translate into large functional improvements and reduced demands on the health care system. These benefits will have a significant impact on her life-long health and wellbeing in a way that medication alone cannot.

REFERENCES

Allen C, Glasziou P, Del Mar C (1999) Bed rest: a potentially harmful treatment needing more careful evaluation. *Lancet* **354**:1229–1233.

Ambrosino N, Bruletti G, Scala V, Porta R, Vitacca M (2002) Cognitive and perceived health status in patients with chronic obstructive pulmonary disease surviving acute or chronic respiratory failure: a controlled study. *Intensive Care Medicine* **28**:170–7.

American College of Sports Medicine (2006) ACSM's guidelines for exercise testing and prescription (7 edn). Pennsylvania: Lippincott Williams & Wilkins.

American Diabetes Association. http://www.diabetes.org/type-2-diabetes/treatment-conditions.jsp Accessed March 2006.

American Heart Association. Blood pressure guidelines. http://www.americanheart.org Accessed March 2006.

Barach AL, Beck GJ (1954) The ventilatory effect of the head-down position in pulmonary emphysema. *American Journal of Medicine* **16**: 55–60.

Behrakis PK, Baydur A, Jaeger MJ, Milic-Emili J (1983) Lung mechanics in sitting and horizontal body positions. *Chest* **83**: 643–646.

Blomqvist CG, Stone HL (1963) Cardiovascular adjustments to gravitational stress. In: Shepherd JT, Abboud FM (eds) *Handbook of Physiology Section 2 Circulation Vol. 2* Maryland: Betheda.

Bolton CF (2001) Critical illness polyneuropathy and myopathy. *Critical Care Medicine* **29**: 2388–2390.

Borg G (1998) *Borg's Perceived Exertion and Pain Scales*. Champaign, Illinois: Human Kinetics.

Bott J, Carroll MP, Conway JH, Keilty SE, Ward EM, Brown AM et al. (1993) Randomised controlled trial of nasal ventilation in acute ventilatory failure due to chronic obstructive airways disease. *Lancet* **341**: 1555–1557.

Braith RW, Vincent KR (1999) Resistance exercise in the elderly person with cardiovascular disease. *American Journal of Geriatric Cardiology* **8**: 63–70.

Burns JR, Jones FL (1975) Letter: Early ambulation of patients requiring ventilatory assistance. *Chest* **68**: 608.

NHS Centre for Reviews and Dissemination, University of York (1998) Cardiac rehabilitation. *Effective Health Care* **4**: 1–12.

Carson SS, Bach PB (2002) The epidemiology and costs of chronic critical illness. *Critical Care Clinics* **18**: 461–476.

Chatte G, Sab JM, Dubois JM, Sirodot M, Gaussorgues P, Robert D (1997) Prone position in mechanically ventilated patients with severe acute respiratory failure. *American Journal of Respiratory and Critical Care Medicine* **155**: 473–478.

Clauss RH, Scalabrini BY, Ray III JF, Reed GE (1968) Effects of changing body position upon improved ventilation-perfusion relationships. *Circulation* **37**: II214–II217.

Convertino VA (1992) Effects of exercise and inactivity on intravascular volume and cardio-vascular control mechanisms. *Acta Astronautica* **27**: 123–129.

Craig DB, Wahba WM, Don HF, Couture JG, Becklake MR (1971) 'Closing volume' and its relationship to gas exchange in seated and supine positions. *Journal of Applied Physiology* **31**: 717–721.

Dantzker DR (1991) Oxygen delivery and utilization. *Applied Cardiopulmonary Pathophysiology* **3**: 345–350.

Dantzker DR, Foresman B, Gutierrez G (1991) Oxygen supply and utilization relationships. A re-evaluation. *American Review of Respiratory Disease* **143**: 675–679.

Dean E (1985) Effect of body position on pulmonary function. *Physical Therapy* **65**: 613–618.

Dean E (1994) Oxygen transport: a physiologically-based conceptual framework for the practice of cardiopulmonary physiotherapy. *Physiotherapy* **80**: 347–353.

Dean E (2006a) Epidemiology as a basis for contemporary physical therapy practice. In: Frownfelter DL, Dean E (eds) *Cardiovascular and Pulmonary Physical Therapy: Evidence and Practice*. St Louis, Missouri: Elsevier.

Dean E (2006b) Mobilization and exercise. In: Frownfelter DL, Dean E (eds) *Cardiovascular and Pulmonary Physical Therapy: Evidence and Practice*. St Louis, Missouri: Elsevier.

Dean E (2006c) Body positioning. In: Frownfelter DL, Dean E (eds) *Cardiovascular and Pulmonary Physical Therapy: Evidence and Practice*. St Louis, Missouri: Elsevier.

Dean E, Perme C (2006) Intensive care unit management of cardiopulmonary dysfunction. In: Frownfelter DL, Dean E (eds) *Cardiovascular and Pulmonary Physical Therapy: Evidence and Practice*. St Louis, Missouri: Elsevier.

Dean E, Ross J (1992) Oxygen transport. The basis for contemporary cardiopulmonary physical therapy and its optimization with body positioning and mobilization. *Physical Therapy Practice* **1**: 34–44.

De Troyer A (1983) Mechanical role of the abdominal muscles in relation to posture. *Respiratory Physiology* **53**: 341–353.

Doering LV (1993) The effect of positioning on hemodynamics and gas exchange in the critically ill: a review. *American Journal of Critical Care* **2**: 208–216.

Don HF, Craig DB, Wahba WM, Couture JG (1971) The measurement of gas trapped in the lungs at functional residual capacity and the effects of posture. *Anesthesiology* **35**: 582–590.

Douglas WW, Rehder K, Beynen FM, Sessler AD, Marsh HM (1977) Improved oxygenation in patients with acute respiratory failure: the prone position. *American Review of Respiratory Disease* **115**: 559–566.

Druz WS, Sharp JT (1981) Activity of respiratory muscles in upright and recumbent humans. *Journal of Applied Physiology* **51**: 1552–1561.

Dubin D (2000) *Rapid Interpretation of EKGs*. Florida: Cover Publishing Company.

Expert Panel (1998) Executive summary of the clinical guidelines on the identification, evaluation, and treatment of overweight and obesity in adults. *Archives of Internal Medicine* **158**: 1855–1867.

Gauer OH, Thron HL (1965) Postural changes in the circulation. In: Hamilton WF (ed.) *Handbook of Physiology*. Washington: American Physiology Society, pp. 439–479.

Gillespie DJ, Rehder K (1987) Body position and ventilation-perfusion relationships in unilateral pulmonary disease. *Chest* **91**: 75–79.

Gosselink RA, Wagenaar RC, Sargeant AJ, Rijswijk H, Decramer MLA (1995) Diaphragmatic breathing reduces efficiency of breathing in chronic obstructive pulmonary disease. *American Journal of Respiratory and Critical Care Medicine* **151**: 1136–1142.

Harvard University School of Public Health. Health risk assessment. http://www.yourdiseaserisk.harvard.edu/ Accessed March 2006.

Heaton KW (1999) Dangers of bed rest. *Lancet* **354**: 2004.

Holten K (1972) Training effect in patients with severe ventilatory failure. *Scandinavian Journal of Respiratory Diseases* **53**: 65–76.

Hoffhuis J, Hautvast JLA, Schrijvers AJP, Bakker J (2003) Quality of life on admission to the intensive care: we query the relatives? *Intensive Care Medicine* **29**: 974–979.

Hofestede G (1980) *Culture's Consequence: International Differences in Work-Related Values*. Beverley Hills, California: Sage.

Hoppeler H, Fluck M (2003) Plasticity of skeletal muscle mitochondria: structure and function. *Medicine and Science in Sports and Exercise* **35**: 95–104.

Hsu HO, Hickey RF (1976). Effect of posture on functional residual capacity postoperatively. *Anesthesiology* **44**: 520–521.

Hu G, Barengo NC, Tuomilehto J, Lakka TA, Nissinen A, Jousilahti P (2004) Relationship of physical activity and body mass index to the risk of hypertension: a prospective study in Finland. *Hypertension* **43**: 25–30.

Ibanez J, Raurich JM, Abizanda R, Claramonte R, Ibanez P, Bergada J (1981) The effect of lateral positions on gas exchange in patients with unilateral lung disease during mechanical ventilation. *Intensive Care Medicine* **7**: 231–234.

Janssen I, Katzmarzyk PT, Ross R (2004) Waist circumference and not body mass index explains obesity-related health risk. *American Journal of Clinical Nutrition* **79**: 379–84.

Jones AYM, Dean E (2004) Body position changes and its effect on hemodynamic and metabolic status. *Heart Lung* **33**: 281–290.

Jones AYM, Dean E, Chow CCS (2003) Comparison of the oxygen cost of breathing exercises and spontaneous breathing in patients with stable chronic obstructive pulmonary disease. *Physical Therapy* **83**: 424–31.

Jones AYM, Yu WC, Mok NS, Yeung OYY, Cheng HCW, Dean E (2006) Exercise-induced desaturation and ECG changes in people with severe lung disease: an exploratory investigation of 25 serial cases. *Heart Lung* (In press).

Jones C, Macmillan RR, Griffiths RD (1994) Providing psychological support for patients after critical illness. *Clinical Intensive Care* **5**: 176–179.

Kaneko K, Milic-Emili J, Dolovich MB, Dawson A, Bates DV (1966) Regional distribution of ventilation and perfusion as a function of body position. *Journal of Applied Physiology* **21**: 767–777.

Kim MJ, Hwang HJ, Song HH (2002) A randomized trial on the effects of body positions on lung function with acute respiratory failure patients. *International Journal of Nursing Studies* **39**: 549–55.

Kollef MH (1999) The prevention of ventilator-associated pneumonia. *New England Journal of Medicine* **340**: 627–634.

Kramer N, Meyer TJ, Meharg J, Cece RD, Hill NS (1995) Randomized, prospective trial of noninvasive positive pressure ventilation in acute respiratory failure. *American Journal of Respiratory and Critical Care Medicine* **151**: 1799–1806.

Lange RA, Katz J, McBride W, Moore Jr DM, Hillis LD (1988) Effects of supine and lateral positions on cardiac output and intracardiac pressures. *American Journal of Cardiology* **62**: 330–333.

Langer M, Mascheroni D, Marcolin R, Gattinoni L (1988) The prone position in ARDS patients. *Chest* **94**: 103–107.

Langou RA, Wolfson S, Olson EG, Cohen LS (1977) Effects of orthostatic postural changes on myocardial oxygen demands. *American Journal of Cardiology* **39**: 418–421.

Leblanc P, Ruff F, Milic-Emili J (1970) Effects of age and body position on 'airway closure' in man. *Journal of Applied Physiology* **28**: 448–451.

Lennon S, Quindry JC, Hamilton KL, French J (2004) Loss of exercise-induced cardioprotection following cessation of exercise. *Journal of Applied Physiology* **96**: 1299–1305.

Levine SA, Lown B (1952) 'Armchair' treatment of acute coronary thrombosis. *Journal of the American Medical Association* **148**: 1365–1368.

MacIntyre NR, Cook DJ, Ely Jr EW, Epstein SK, Fink JB, Heffner JE et al. (1977) Respiratory muscle fatigue: A cause of respiratory failure? *Clinical Science & Molecular Medicine* **53**: 419–422.

McGavin CR, Artvinli M, Naoe H, McHardy GJR (1978) Dyspnea, disability, and distance walked: Comparison of estimates of exercise performance in respiratory disease. *British Medical Journal* **2**: 241–243.

Mellins RB (1974) Pulmonary physiotherapy in the pediatric age group. *American Review of Respiratory Diseases* **110**; Suppl. 1: 137–142.

Mikines KJ, Richter EA, Dela F, Galbo H (1991) Seven days of bed rest decrease insulin action on glucose uptake in leg and whole body. *Journal of Applied Physiology* **70**: 1245–1254.

New York Heart Association (NYHA) Classification for Congestive Heart Failure (CHF). http://www.hcoa.org/hcoacme/chf-cme/chf00070.htm Accessed March 2006.

Noonan V, Dean E (2000) Submaximal exercise testing: clinical application and interpretation. *Physical Therapy* **80**: 782–807.

Oken BS, Zajdel D, Kishiyama S, Flegal K, Dehen C, Haas M et al. (2006) Randomized controlled six-month trial of yoga in health seniors: effects on cognition and quality of life. *Alternative Therapies in Health and Medicine* **12**: 40–7.

Ornish D (1998) Avoiding revascularization with lifestyle changes: the Multicenter Lifestyle Demonstration Project. *American Journal of Cardiology* **82**: 72T–76T.

Ornish D, Scherwitz LW, Billings JH, Brown SE, Gould KL, Merritt TA et al. (1998) Intensive lifestyle change for reversal of coronary heart disease. *Journal of the American Medical Association* **280**: 2001–2007.

Peruzzi WT (2005) Sleep in the intensive care unit. *Pharmacotherapy* **25**: 34S–9S.

Piehl MA, Brown RS (1976) Use of extreme position changes in acute respiratory failure. *Critical Care Medicine* **4**:13–14.

Plant PK, Owen JL, Parrott S, Elliott MW (2003) Cost effectiveness of ward based non-invasive ventilation for acute exacerbations of chronic obstructive pulmonary disease: economic analysis of randomised controlled trial. *British Medical Journal* **326**: 956–60.

Prakash RW, Parmley W, Dikshit K, Forrester J, Swan HK (1973) Hemodynamic effects of postural changes in patients with acute myocardial infarction. *Chest* **64**: 7–9.

Prefaut C, Engel LA (1981) Vertical distribution of perfusion and inspired gas in supine man. *Respiratory Physiology* **43**: 209–219.

Probst VS, Heyvaert H, Coosemans I, Pitta F, Spruit MA, Troosters T et al. (2003) Effects of a rollator on exercise capacity, gas exchange and ventilation in COPD patients. *American Journal of Respiratory and Critical Care Medicine* **167**: A669.

Prochaska JO, DiClemente CC (1982) Transtheoretical therapy: toward a more integrative model of change. *Psychotherapy: Theory, Research and Practice* **9**: 276–288.

Ram FS, Lightowler JV, Wedzicha JA (2003) Non-invasive positive pressure ventilation for treatment of respiratory failure due to exacerbations of chronic obstructive pulmonary disease. *Cochrane Library* http://www.thecochranelibrary.com CD004104.

Ray III JF, Yost L, Moallem S, Sanoudos GM, Villamena P, Paredes RM et al. (1974) Immobility, hypoxemia, and pulmonary arteriovenous shunting. *Archives of Surgery* **109**: 537–541.

Remolina C (1981) Positional hypoxaemia in unilateral lung disease. *New England Journal of Medicine* **304**: 523–525.

Roussos CH, Fukuchi Y, Macklem PT, Engel LA (1976). Influence of diaphragmatic contraction on ventilation distribution in horizontal man. *Journal of Applied Physiology* **40**: 417–424.

Saltin B, Blomqvist G, Mitchell JH, Johnson BL (1998) Response to exercise after bed rest and after training. *Circulation* **38**(VII): S1–S78.

Sandler H (1986) Cardiovascular effects of inactivity. In: Sander H and Vernikos J (eds) *Inactivity: its physiological effects* New York: Academic Press Inc, pp. 11–47.

Sasaki H, Hida W, Takishima T (1977) Influence of body position on dynamic compliance in young subjects. *Journal of Applied Physiology* **42**: 706–710.

Sato Y (2000) Diabetes and life-styles: role of physical exercise for primary prevention. *British Journal of Nutrition* **84**; Suppl.: S187–90.

Sharp JT, Druz WS, Moisan T, Foster J, Machnach W (1980) Postural relief of dyspnea in severe chronic obstructive pulmonary disease. *American Review of Respiratory Disease* **122**: 201–211.

Shichinohe Y, Ujike Y, Kurihara M, Yamamoto S, Oota K, Tsukamoto M et al. (1991) [Respiratory care with prone position for diffuse atelectasis in critically ill patients]. *Kokyu To Junkan* **39**: 51–55.

Shoemaker WC (ed.) (1999) *Textbook of Critical Care* (4 edn) Philadelphia: Elsevier.

Short Form-36 Questionnaire http://www.swin.edu.au/victims/resources/assessment/health/sf-36-questionnaire.html Accessed March 2006.

Singelis TM, Triandis HC, Bhawuk DPS, Gelfand M (1995) Horizontal and vertical dimensions of individualism and collectivism: a theoretical measurement refinement. *Cross Cultural Research* **29**: 240–75.

Sjostrand T (1951) Determination of changes in the intrathoracic blood volume in man. *Acta Physiologica Scandinavica* **22**:116–128.

Sonneblick M, Meltzer E, Rosin AJ (1983) Body positional effect on gas exchange in unilateral pleural effusion. *Chest* **83**: 784–786.

Svanberg L (1957) Influence of posture on lung volumes, ventilation and circulation of normals. *Scandinavian Journal of Clinical Laboratory Investigation* **25**: 1–195.

Tudor-Locke C, Bassett Jr DR (2004) How many steps/day are enough? Preliminary pedometer indices for public health. *Sports Medicine* **34**: 1–8.

Wagaman MJ, Shutaack JG, Moomjiam AS, Schwartz JG, Shaffer TH, Fox WW (1979) Improved oxygenation and lung compliance with prone positioning of neonates. *Journal of Pediatrics* **94**: 787–791.

Walder B, Francioli D, Meyer JJ, Lancon M, Romand JA (2000) Effects of guidelines implementation in a surgical intensive care unit to control nighttime light and noise levels. *Critical Care Medicine* **28**: 2242–2247.

Wasserman KJ, Hansen E, Sue DY, Whipp BJ, Casaburi R (1994) *Principles of Exercise Testing and Interpretation* (2 edn) Philadelphia: Lea & Febiger.

Weissman C, Kemper M (1993) Stressing the critically ill patient: the cardiopulmonary and metabolic responses to an acute increase in oxygen consumption. *Journal of Critical Care* **8**: 100 Lea & Febiger, 108.

Weissman C, Kemper M, Damask MC, Askanazi J, Hyman AI, Kinney JM (1984) Effect of routine intensive care interactions on metabolic rate. *Chest* **86**: 815–818.

Weissman C, Kemper M, Elwyn DH, Askanazi J, Hyman AI, Kinney JM (1989) The energy expenditure of the mechanically ventilated critically ill patient. *Chest* **2**: 254–259.

Weissman C, Kemper M, Harding J (1994) Response of critically ill patients to increased oxygen demand: hemodynamic subsets. *Critical Care Medicine* **22**: 1809–1816.

Wenger NK (1982) Early ambulation: the physiologic basis revisited. *Advances in Cardiology* **31**: 138–141.

Wong WP (2000) Physical therapy for a patient in acute respiratory failure. *Physical Therapy* **80**: 662–70.

World Health Organisation (1997) Consultation on Obesity, Geneve, Switzerland.

World Health Organisation (2002) International Classification of Functioning, Disability and Health. http://www.sustainable-design.ie/arch/ICIDH-2PFDec-2000.pdf Accessed March 2006.

Yohannes AM, Connolly MJ (2003) Early mobilization with walking aids following hospital admission with acute exacerbation of chronic obstructive pulmonary disease. *Clinical Rehabilitation* **17**: 465–471.

Zack MB, Pontoppidan H, Kazemi H (1974) The effect of lateral positions on gas exchange in pulmonary disease. A prospective evaluation. *American Review of Respiratory Disease* **110**: 49–55.

II Surgical

3 Abdominal Surgery: The Evidence for Physiotherapy Intervention

LINDA DENEHY AND LAURA BROWNING

OVERVIEW

The objective of this chapter is to present the evidence for the physiotherapy management of a patient having major abdominal surgery using a case based scenario. The evidence for the physiotherapy management of Mr C, a 69 year old male undergoing upper abdominal surgery (UAS) presented in the case below, will be discussed by posing seven important clinical questions. Upper abdominal surgery is defined as surgery involving 'an incision above or extending above the umbilicus' (Celli et al. 1984 **A**). This case represents a common scenario encountered on surgical wards in public and larger private hospitals worldwide.

CASE REPORT

69 year old male.
Presented to the out-patient clinic with rectal bleeding and loss of weight (7kg in 2 months).
Investigations revealed colon cancer located at the hepatic flexure.

PAST MEDICAL HISTORY

Mild Chronic Obstructive Pulmonary Disease diagnosed three years ago.
Rectal polyps.
Gout.

SOCIAL HISTORY

Ex smoker – previously smoked one packet of cigarettes daily for 45 years, quit three years ago.
Lives with wife in two-storey home.
Retired bank manager.
Social drinker.
Plays golf twice weekly.

Recent Advances in Physiotherapy. Edited by C. Partridge
© 2007 John Wiley & Sons, Ltd

PHYSICAL FUNCTION

Shortness of breath climbing hills and stairs.
Exercise tolerance approximately 1 km.
Distance reduced with exacerbations of gout.
Nil gait aids required.

RESPIRATORY HISTORY

Morning cough with small amounts of white sputum.
FEV_1/FVC: 65 %.
FEV_1:67 % predicted.
CXR: hyper inflated lung fields, no focal consolidation.

MEDICATIONS

Tiotropium Bromide 10 mcg via Handihaler once daily.
Salbutamol MDI as required.
Symbicort Turbuhaler once daily.

PRE-OPERATIVE ASSESSMENT

BMI 28.5.
Slightly barrel shaped chest and reduced chest expansion.
Reduced breath sounds with occasional expiratory wheeze.
Strong, dry, non-productive cough.
Oxyhaemoglobin Saturation (SpO_2) 96 % on room air.

OPERATIVE HISTORY

Extended right hemicolectomy via midline laparotomy.
Anaesthetic duration 180 minutes.
American Society of Anaesthetists Score 3 (American Society of Anaesthesiologists
    1963).
On return from theatre: stable condition, temperature 37.2 °C, pulse rate 90, blood
    pressure 110/70, oxygen therapy via Hudson mask 6 L/minute, $SpO_2 = 96$ %.
Analgesia: morphine PCA 1mg with five minute lockout interval.

INTRODUCTION

Post-operative pulmonary complications (PPC) were first identified as early as 1910
by Pasteur, who postulated that active collapse of the lung resulted from a deficiency
of respiratory power (Pasteur 1910 C). Perioperative physiotherapy treatment has
played a significant role in minimising the adverse effects of anaesthesia and surgery

on the respiratory system for more than 50 years. The physiotherapy techniques applied during treatment aim to counteract adverse pulmonary changes such as low lung volumes, atelectasis and secretion retention (Stiller et al. 1994 **A**).

Recent advances in both surgical and pain management, the evolution of new forms of perioperative physiotherapy techniques and a reduction in the incidence of clinically significant PPC have provided the stimuli for a re-evaluation of the role of physiotherapy in all forms of major surgery. In most Western hospitals physiotherapy services are provided to major surgical units, where physiotherapists commonly treat patients both pre- and post-operatively. However, service provision varies widely in response to external influences such as surgeon preferences and implementation of change to practice in response to recent or new evidence.

QUESTION 1

Will provision of physiotherapy treatment for Mr C reduce his risk of developing a post-operative pulmonary complication?

There are three strands of knowledge it is necessary to consider in answering this question: basic science, published evidence from high quality clinical trials, and knowledge generated from professional practice (Herbert et al. 2005 **C**). The first two will be discussed in detail and the third only briefly.

BASIC SCIENCE

Alterations in pulmonary function are an expected intraoperative and post-operative finding, especially following UAS (Craig 1981 **R**; Durreuil et al. 1987 **A**; Ford et al. 1993 **R**). The characteristic post-operative abnormality is a restrictive ventilatory pattern with reductions in vital capacity (VC) and functional residual capacity (FRC) (Meyers et al. 1975 **A**). The post-operative breathing pattern is shallow, with an increased rate of respiration (Duggan & Kavanagh 2005 **R**). Reductions in FRC have been demonstrated immediately upon induction of general anaesthesia (Wahba 1991 **R**) and may affect airway calibre, airway closure, lung compliance and gas exchange, leading to atelectasis (Nunn 1990 **R**). The relationship of FRC with the closing capacity of the lungs (CC) explains the significance of perioperative reductions in FRC. If the CC exceeds the FRC then dependent lung regions under-ventilate, resulting in ventilation/perfusion mismatch and hypoxaemia (Craig 1981 **R**). Closing capacity increases with the loss of elastic lung tissue that occurs with increasing age and in chronic lung disease (Fairshter & Williams 1987 **R**). In combination with these increases in CC, any factor which at the same time reduces FRC will significantly affect the relationship between the two volumes, such that dependent airway closure occurs, resulting in atelectasis. Furthermore, mucociliary clearance is adversely affected by the reduction in lung volumes, causing reduced cough effectiveness. During surgery, the introduction of anaesthetic gases also impairs mucociliary clearance by depressing mucociliary flow (Konrad et al. 1995 **B**).

In view of the physiological changes occurring in the respiratory system as a result of UAS, two basic theories have been proposed to explain the pathogenesis of PPC: regional hypoventilation and blockage of airways by mucus. Advocates of the mucus blockade theory contend that the primary cause of atelectasis is the absorption of alveolar air distal to a mucus plug in the proximal airway, causing eventual collapse unless fresh air enters through collateral channels (Gamsu et al. 1976 **R**; Marini 1984 **R**). These authors, and others (Forbes 1976 **A**; Lansing & Jamieson 1963 **R**) suggest that the cumulative effect of the perioperative process presents a significant insult to mucociliary clearance. The second basic process thought to cause PPC is regional hypoventilation. There are several physiological factors which may contribute to alveolar closure; these relate to reductions in FRC and an altered relationship between FRC and CC, together with marked diaphragmatic dysfunction post-operatively as discussed above. The precise sequence and relative contribution of each of the two mechanisms for developing PPC is still unclear. Other risk factors which may predispose to increased risk of mucus plugging are a history of smoking, weak cough, prolonged intubation, presence of a nasogastric tube and prolonged post-operative atelectasis (Smith & Ellis 2000 **R**).

The physiological changes occurring in the lungs after major surgery and the proposed theories of pathogenesis of PPC provide empirical support for the use of physiotherapy intervention to counteract these changes and reduce Mr C's risks of developing a PPC. Further support is provided by randomised controlled trials that compare physiotherapy treatment as a total entity to no treatment.

EVIDENCE FROM CLINICAL TRIALS

An extensive database search of the literature was undertaken using MEDLINE, CINAHL, ISI Web of Science, PEDro and Evidence Based Medicine Reviews (Cochrane, DARE). The search terms entered included 'pulmonary complications', 'atelectasis', 'pneumonia and surgery', 'respiratory therapy', 'chest physiotherapy', 'chest physical therapy', 'breathing exercises', 'early mobilisation', 'early ambulation', 'continuous positive airway pressure' (CPAP), 'incentive spirometry' (IS) and 'positive expiratory pressure' (PEP).

Six randomised controlled trials provide level 1b to 2b evidence (Sackett et al. 2000 **C**) for the effectiveness of physiotherapy in preventing PPC following UAS when compared to no treatment. A summary of these trials is given in Table 3.1. The methodological quality of each of the trials was assessed using the PEDro scale (Centre for Evidence-Based Physiotherapy 2006 **C**). Absolute risk reduction (including confidence intervals) and number needed to treat (NNT) have been calculated from the dichotomous PPC data supplied in the articles (Herbert 2000 **R**). Five of the trials provide moderate quality evidence (Celli et al. 1984 **A**; Chumillas et al. 1998 **A**; Condie et al. 1993 **A**; Morran et al. 1983 **A**; Olsen et al. 1997 **A**) with a PEDro score greater than 5/10. Only three of these trials present convincing evidence that

Table 3.1. Summary of randomised controlled trials comparing physiotherapy to no treatment following UAS

Author/ Year	Sample Size	Intervention	PPC Definition	PPC Incidence	Conclusion	ARR (95 % CI)	NNT	PEDro scale score
Olsen et al. 1997	368	Treatment: Pre-operative physiotherapy; breathing exercises with pursed lips; huffing and coughing hourly; information about the importance of early mobilisation; PEP for high risk pts. Control: No intervention.	$SpO_2 < 92\%$ or two of: Temp $>38.2\,°C$, pathological lung auscultation, radiological evidence of pneumonia/ atelectasis.	Treatment: 6 % Control: 27 %	Pre-operative chest physiotherapy reduced the incidence of PPC and improved mobilisation and oxygen saturation after major abdominal surgery.	0.21 (0.14 to 0.28)	5	5/10
Condie et al. 1993	130 (310 total, only 130 major UAS)	Treatment: Pre-operative physiotherapy; daily physiotherapy for 3 days post-operatively. Control: Pre-operative physiotherapy; no post-operative supervision, just followed information sheet	One of: Temp $>38\,°C$ with abnormal auscultation findings, temp $>38\,°C$ with abnormal sputum production, or abnormal sputum production alone.	Treatment: 8.2 % Control: 17.4 %	The value of the routine provision of supervised post-operative chest physiotherapy in non-smoking patients undergoing elective abdominal surgery is questionable.	0.09 (−0.03 to 0.21)	11	6/10

(Continued)

Table 3.1. Summary of randomised controlled trials comparing physiotherapy to no treatment following UAS (*Continued*)

Author/ Year	Sample Size	Intervention	PPC Definition	PPC Incidence	Conclusion	ARR (95 % CI)	NNT	PEDro scale score
Chumillas et al. 1998	81	Treatment: Respiratory rehabilitation including FET, DBE, SMI and early mobilisation. Control: No intervention.	Bronchitis: negative CXR, Temp >37.5 °C, sputum abundant and clear. Atelectasis: CXR collapse, Temp >38 °C, diminished breath sounds. Pneumonia: CXR shows consolidation, Temp >38 °C, crackles on auscultation, sputum abundant and purulent.	Treatment: 7.5 % Control: 19.5 %	Respiratory rehabilitation protects against PPC and is more effective in moderate and high risk patients, but does not affect surgery induced functional alterations.	0.12 (−0.03 to 0.27)	8	5/10
Celli et al. 1984	81 (172 total, 81 UAS)	Treatment: Pre-operative physiotherapy; 4 times daily for 4 post-operative days. IPPB: 15 mins IPPB. IS: 10 breaths up to 70 % VC. DBE: 6 × 10 DBE with SMI and cough. Control: No intervention.	3 or more of: cough, sputum, dyspnoea, chest pain, fever >38 °C, tachycardia.	IPPB: 30.4 % IS: 33.3 % DBE: 33.3 % Control: 89.5 %	IPPB, IS and DBE, when compared to an untreated control group, were equally effective in significantly decreasing the incidence of PPC after abdominal surgery.	IPPB vs control: 0.59 (0.30 to 0.76) IS vs control: 0.56 (0.26 to 0.74) DBE vs control: 0.56 (0.25 to 0.75)	IPPB vs control: 2 IS vs control: 2 DBE vs control: 2	6/10

(*Continued*)

Table 3.1. Summary of randomised controlled trials comparing physiotherapy to no treatment following UAS (*Continued*)

Author/ Year	Sample Size	Intervention	PPC Definition	PPC Incidence	Conclusion	ARR (95 % CI)	NNT	PEDro scale score
Roukema et al. 1988	153	Treatment: Pre-operative physiotherapy; post-operative DBE, FET and coughing, and dead space rebreathing. Control: No intervention.	3 grades: 1 minor atelectasis, no hypoxaemia, no fever. 2 minor atelectasis, hypoxaemia, no fever. 3 major atelectasis, hypoxaemia, fever.	All grades combined: Treatment: 19 % Control: 60 % Only grades 2 & 3: Treatment: 4 % Control: 35 %	Pre- and post-operative breathing exercises as a prophylactic treatment in all patients scheduled for UAS are recommended.	Grades 2 & 3: 0.30 (0.18 to 0.41)	Grades 2 & 3: 3	1/10
Morran et al. 1983	102	Treatment: 15 mins DBE, assisted coughing and vibration. Control: No intervention.	Pulmonary atelectasis: pyrexia, production of sputum, clinical and radiological evidence of collapse. Chest infection: pyrexia, production of purulent sputum, clinical signs of infection and radiological signs of collapse.	Chest infection: Treatment: 14 % Control: 37 % Pulmonary atelectasis: Treatment: 22 % Control: 35 % Combined: Control: 59 % Treatment: 49 %	Routine prophylactic post-operative chest physiotherapy decreased significantly the frequency of chest infection.	Combined: 0.10 (−0.09 to 0.28) Chest infection: 0.24 (−0.07 to 0.39)	Combined: 10 Chest infection: 4	5/10

ARR: Absolute risk reduction; CI: Confidence interval; CXR: Chest X-ray; DBE: Deep breathing exercises; FET: Forced expiration technique; IPPB: Intermittent positive pressure breathing; IS: Incentive spirometry; NNT: Number need to treat; PEP: Positive expiratory pressure; PPC: Post-operative pulmonary complication; SMI: Sustained maximal inspiration; Temp: Temperature; VC: Vital capacity.

physiotherapy treatment reduces the incidence of PPC following UAS (Celli et al. 1984 **A**; Morran et al. 1983 **A**; Olsen et al. 1997 **A**).

Morran and co-workers (1983 **A**) monitored breathing exercises, vibration and coughing for two or more days after surgery in 102 subjects, comparing a group receiving physiotherapy treatment with a no treatment group. It was stated that both groups received encouragement from nursing and medical staff to take deep breaths and cough. The primary outcome measure was incidence of PPC. The authors concluded that routine prophylactic physiotherapy reduced the frequency of post-operative chest infection. On average, one respiratory complication was prevented in every four treated patients, that is, the NNT was four.

In this study there was good baseline equivalence between groups, however there is no discussion of the method of randomisation, inclusion or exclusion criteria or patient withdrawals. While the criteria used for diagnosis of PPC were reflective of a clinically relevant complication according to other reports (O'Donohue 1992 **R**), there was no indication of who made the final diagnosis of PPC. Furthermore, there is no reference to the position of the patients during treatment, whether treatments were administered pre-operatively as well as post-operatively or if and when patients were mobilised. In a clinical trial such as this, encouragement from staff for both groups of patients to deep breathe and cough is to be expected and is difficult to control (Morran et al. 1983 **A**). These methodological problems are common to other earlier studies examining the role of physiotherapy in UAS and explain why the PEDro score is 5/10.

The following year, a study by Celli and co-workers (1984 **A**) demonstrated a dramatic reduction in PPC and a reduced hospital length of stay (LOS) in subjects receiving physiotherapy treatment. Subjects were allocated to one of four groups receiving either: intermittent positive pressure breathing (IPPB), IS, deep breathing exercises (DBE), or no intervention. LOS was reduced in all treatment groups, however the only significant reduction occurred in the group receiving IS. The authors concluded that this study supports the use of physiotherapy treatment over no treatment in reducing the incidence of PPC. The definition of PPC was clinically based and the NNT was also four. While the methods of this study were better documented than those of Morran and co-workers (1983 **A**) (it received a PEDro score of 6/10), the authors failed to document the patient mobility level and method of pain management, both potential confounding factors.

A more recent large clinical study of 368 Swedish patients provides strong evidence for the role of physiotherapy in reducing PPC when compared to a no treatment control group (Olsen et al. 1997 **A**). The treatment group received pre- and post-operative physiotherapy consisting of PEP mask therapy. The method was well described and potential confounding variables such as ambulation were measured or controlled. 27 % of patients in the control group, compared with only 7 % in the treatment group, developed a clinically relevant PPC. The NNT was found to be five, with tight confidence intervals.

The remaining three randomised controlled trials do not provide conclusive evidence due to a variety of methodological flaws. Chumillas and colleagues (1998 **A**) in their study of 81 subjects undergoing UAS, reported a difference of 12 % in the incidence of PPC between subjects receiving physiotherapy treatment and controls. While the NNT was eight in this study, the confidence intervals were wide (-4 to 29). This no doubt reflects the fact that the sample size was too small given the relatively low risk of subjects and the low incidence of PPC in their cohort.

Condie and colleagues (1993 **A**) investigated the value of routine provision of post-operative deep breathing exercises, huff and cough in 330 subjects, who were non-smokers undergoing elective UAS. Subjects were described as having a low risk of developing PPC, and those with chronic respiratory disease were excluded. Inspection of the sample characteristics indicates that a large proportion of lower abdominal surgery (LAS) patients were included in the study. The definition of UAS was questionable, with surgery involving an incision with more than 50 % of the wound above the umbilicus being classed as UAS, and surgery with more than 50 % of the wound below the umbilicus being LAS. To improve interpretation of the data, we inspected the specific type of operation performed, rather than the incision classification. When gynaecological surgery and hernia repairs were removed, a sample of 130 subjects remained. Both groups received pre-operative education. The incidence of PPC in the group receiving supervised post-operative physiotherapy was 8.2 %, while in the group not receiving post-operative physiotherapy it was 17.4 %. Although there appears to be a difference in the incidence of PPC, the result was not significant. It is questionable whether the sample size was sufficient to detect significant differences in a low risk sample such as this.

The study by Roukema and co-workers (1988 **A**) was of poor quality (PEDro score of 1/10) and will not be considered in this discussion.

Based on these randomised clinical trials, the provision of physiotherapy treatment to reduce the incidence of PPC in patients having UAS will on average prevent one respiratory complication in every four or five patients treated.

A further variable in the studies under discussion is the subjects' level of risk for the development of PPC. Several patient risk factors have been associated with an increased incidence of PPC, however to date no highly sensitive and specific published risk screening model is available for use by clinicians in the UAS population.

Risks factors associated with the development of PPC have been studied extensively and a summary of the common risk scoring systems found in the studies already discussed is shown in Table 3.2. Chumillas and colleagues (1998 **A**) based their scoring on the work of Torrington and Henderson (1988 **A**). Hall and co-workers (1991 **A**) and Brooks-Brunn (1997 **R**) have also contributed to the body of literature searching for a valid risk factor model; their studies are also included in Table 3.2.

Olsen and colleagues (1997 **A**) found a significantly greater incidence of PPC in the subjects they classified as high risk. In the treatment group 15 %, and in the control group 51 % of subjects defined as high risk developed a PPC. This indicates a NNT of three, compared with five for all patients. Chumillas and co-workers (1998 **A**) found a greater incidence of PPC in their high risk subject group, however there were no significant differences in PPC between control and treatment groups in any of the risk categories. While these results and professional experience support the notion that increased risk may lead to an increased incidence of PPC, the risk assessments used in

Table 3.2. Risk scoring systems presented chronologically

Author/Year	Sample Size	Population Studied	Risk Model	Predictive Ability
Torrington & Henderson 1988	1,476	UAS & LAS	Spirometry Age >65 yr BMI >150 % Surgery location Pulmonary history (Points assigned for each risk, high risk >7 points)	43 %
Hall et al. 1991	1,000	UAS	ASA >1 Age >59 yr (Both criteria present)	88 %
Brooks-Brunn 1997	400	UAS & LAS	Age >60 yr BMI >27 % Impaired cognitive function History of cancer Smoking in last 8 weeks Abdominal incision (System not described)	79 %
Olsen et al. 1997	368	UAS	Age >50 yr Smoking history BMI >30 % Pulmonary disease requiring medication Reduced ventilatory function (Need age plus at least one of other)	51 %
Brooks-Brunn 1998	276	UAS & LAS	Abdominal incision Incision length >30 cm Angina ASA >3 (System not described)	77 %

ASA: American society of anaesthesiologists score; BMI: Body mass index; LAS: Lower abdominal surgery; UAS: Upper abdominal surgery.

these studies have not been validated. The risk assessment model used by Olsen and colleagues (1997 **A**) demonstrated poor sensitivity and predictive ability (sensitivity 38 %, predictive ability 51 %) (Scholes 2005 **A/R**).

To allow clinicians to make appropriate and valid assessments about which risk factors are most predictive of PPC, a sensitive and specific multivariate risk model with good clinical utility is required. In a prospective study of 1,055 subjects, age, a positive cough test, presence of a perioperative nasogastric tube, and duration of anaesthesia greater than 2.5 hours were found to be independently associated with increased risk of PPC (McAlister et al. 2005 **A**). Scholes and colleagues (2006 **A**) developed and piloted a weighted risk prediction model for the development of PPC in 272 subjects undergoing UAS. The risk factors producing the most sensitive and specific model were: surgical category, anaesthetic duration greater than 180 minutes, presence of respiratory co-morbidity, history of smoking, and self-reported $V0_2$ maximum (determined by administering a short functional questionnaire outlined by Rankin et al. (1996 **A**)). The model predicted 82 % of subjects who developed a PPC and found that high risk subjects were 8.4 times more likely to develop a PPC than low risk subjects. This model is superior in its predictive ability to the previously published models summarised in Table 3.2. Assessment of the risk of developing PPC is important for the physiotherapist as it allows prioritised respiratory care for high risk subjects and more appropriate use of scarce resources in physiotherapy staffing.

Applying the basic premise of this model (without the calculations) and that of McAlister and co-workers (2005 **A**), Mr C possesses several of the risk factors as outlined below and therefore should be considered a high risk candidate for developing PPC:

Duration of surgery: 180 minutes
History of smoking
Respiratory co-morbidity
UAS – colorectal surgery

Does This Justify Treatment of Mr C?

Three randomised controlled trials of moderate quality (PEDro score >5) examining a total of 451 subjects provide evidence supporting the treatment of Mr C. Only a small proportion of physiotherapy clinical trials are of high quality as blinding of subjects and therapists is often impossible, therefore a PEDro score of five represents the standard quality of published trials (Herbert et al. 2005 **C**). The three trials presented were conducted between 1983 and 1997 and represent subjects and procedures from three different Western countries (Sweden, the United States and the United Kingdom). The demographic profile and treatment methods used in these trials broadly relate to the case of Mr C.

Four published literature reviews – three systematic and one narrative – provide further evidence to support the use of physiotherapy techniques in preventing PPC following UAS (Lawrence et al. 2005 **R**; Overend et al. 2001 **R**; Thomas & McIntosh 1994 **R**; and Olsen 2000 R). A summary of these reviews is provided in Table 3.3. Systematic reviews provide combined evidence from several trials using systematic and explicit methodology and therefore offer the highest level of evidence, whereas narrative reviews may introduce bias in interpretation.

In a recently published abstract, Lawrence and colleagues (2005 **R**) systematically reviewed the effects of surgical, medical and physiotherapy interventions on PPC prevention in non-cardiopulmonary surgery. The results of this review indicate that

Table 3.3. Published literature reviews examining the role of prophylactic physiotherapy in reducing PPC following UAS

Author/Year	Type of Review	Aims	Techniques Examined	Conclusion
Overend et al. 2001	Systematic	To systematically review the evidence examining the use of IS for the prevention of PPC.	IS	Presently, the evidence does not support the use of IS for decreasing the incidence of PPCs following cardiac surgery or UAS.
Thomas & McIntosh 1994	Systematic (and meta-analysis)	To quantitatively assess the conflicting bodies of literature concerning the efficacy of IS, IPPB and DBE in the prevention of PPC in patients undergoing UAS.	IS IPPB DBE	IS and DBE appear more effective than no physical therapy intervention in the prevention of PPC. There is no evidence to support a significant difference between any of the three modalities.
Lawrence et al. 2005	Systematic	To perform a systematic review of the evidence for interventions to prevent PPC after non-cardiopulmonary surgery.	Postoperative lung expansion (IS, DBE), Nasogastric decompression, Smoking cessation prior to surgery, Short-acting neuromuscular blockade, Immune-enhancing enteral formulations, Epidural analgesia, Incision type.	Proven interventions to reduce PPC include post-operative lung expansion and selective nasogastric decompression.
Olsen 2000	Narrative	To review studies on the effects of chest physiotherapy in open abdominal surgery.	IS IPPB DBE PEP Postural drainage	The results showed the beneficial effects of DBE in preventing PPC, especially in high risk patients.

DBE: Deep breathing exercises; IPPB: Intermittent positive pressure breathing; IS: Incentive spirometry; PEP: Positive expiratory pressure; PPC: Post-operative pulmonary complication.

there is good evidence to support the use of physiotherapy techniques in preventing PPC. The remaining reviews will be discussed later in this chapter.

A group of experienced cardiorespiratory physiotherapists working in Australia indicated that prevention of one PPC in every 20 treatments was the minimal clinically worthwhile number (Herbert 2000 **R**). This represents a NNT of 20.

On the basis of this research, the evidence from clinical trials supports the role of physiotherapy treatment for Mr C. Additional knowledge of risk factors for developing PPC further supports the provision of physiotherapy treatment in this case.

PROFESSIONAL PRACTICE

The cost effectiveness of providing physiotherapy resources to UAS populations based upon this philosophy has not yet been documented. Additionally, the influence of other contextual factors such as cultural influences and hospital policies are not considered in this evidence (Herbert et al. 2005 **C**). The influence of professional practice knowledge gained from reflection on the day to day treatment of similar patients assists physiotherapists in their complex clinical reasoning processes. This knowledge allows physiotherapists to integrate patient preferences and professional and basic science knowledge with evidence from clinical trials to ensure treatment is relevant to each particular clinical situation (Herbert et al. 2005 **C**). It is this third strand of knowledge that integrates the clinical decision making processes with the available evidence to ensure the most appropriate evidence-based decisions are reached.

QUESTION 2

How are post-operative pulmonary complications defined in the literature?

Despite a significant volume of research, the precise definition of a PPC, its causative factors and the true incidence of PPC in surgical populations remain unknown. Pulmonary complications documented in the literature include atelectasis, hypoxaemia and pneumonia (Ali et al. 1974; Brooks-Brunn 1995b **R**; Craig 1981 **R**). Less commonly, pulmonary embolus, pleural effusion and pneumothorax are reported (Ridley 1998 **C**). Of these, pulmonary atelectasis is the most commonly reported respiratory complication (O'Donohue 1985 **C**).

The incidence of PPC is a function of the diagnostic criteria used. As a result of the differing criteria used to define a PPC and failure to further identify a clinically significant PPC, the incidence reported in the literature varies considerably. It has been reported to be between 5 and 75 % (Dilworth & White 1992 **A**). The incidence of atelectasis measured using chest radiography has been reported to be approximately 70 %, however, clinically significant PPC develop in few of these patients (Denehy 2002 **A/R**; Jenkins et al. 1990 **A**; O'Donohue 1985 **C**). Bourn and Jenkins (1992 **R**) describe post-operative atelectasis as 'the rule rather than the exception'; this view is supported in other literature (Platell & Hall 1997 **R**). In more recent studies where a multi-criteria outcome has been used, the reported incidence of PPC was as low as

Table 3.4. An example of a definition of a clinically significant PPC using multiple
outcome measures (Scholes 2005 A/R)

PPC diagnosis was confirmed when four or more of the following signs and symptoms
were present:

- Chest radiograph report of collapse/consolidation.
- Raised temperature $>38\,°C$ on two or more consecutive days.
- SpO_2 $<90\,\%$ on room air on two consecutive days.
- Production of yellow or green sputum which is different to any in pre-operative
  assessment.
- An otherwise unexplained white cell count $> 11 \times 10^9/L$, or prescription of an
  antibiotic specific for respiratory infection.
- Physician diagnosis of chest infection.
- Presence of infection on sputum culture report.
- Abnormal breath sounds on auscultation which differ from any in pre-operative
  assessment.

5–20 % (Brooks-Brunn 1997 **A**; Hall et al. 1996b **A**; Jenkins et al. 1989 **A**; Mackay
et al. 2005 **A**; Stiller et al. 1995 **A**).

More recent papers attempt to define a PPC with reference to the clinical significance of the problem, which includes consideration of both hospital and patient costs.
O'Donohue (1992 **R**) defines a PPC as 'a pulmonary abnormality that produces identifiable disease or dysfunction that is clinically significant and adversely affects the
clinical course'. However, specific outcome criteria which accurately describe clinically relevant complications remain elusive. Studies using a combination of multiple
outcome measures rather than single variables may more accurately define a clinically
significant PPC. The definition of a PPC employed in research conducted by Scholes
(2005 **A/R**) provides an example of this and is displayed in Table 3.4.

Having established that physiotherapy treatment of Mr C is required, a further
clinical question arises.

QUESTION 3

Which physiotherapy technique is most effective in reducing the risk of PPC?

There are several well recognised physiotherapy techniques employed in the treatment of patients undergoing UAS. A large body of literature exists comparing the
efficacy of one technique with another in the prevention of PPC. The majority of early
research conducted poorly controls for confounding variables such as patient mobilisation, adherence to treatment protocols and pain levels (Hallbook et al. 1984 **A**;
Thomas & McIntosh 1994 **R**).

The physiotherapy techniques examined in the literature include pre-operative education, deep breathing strategies, IS, PEP, CPAP, IPPB and early mobilisation. A
summary of the research is presented in Table 3.5. It can be seen from the variable
results of these comparative studies, which generally provide level 3 evidence

Table 3.5. Comparative physiotherapy research in patients following UAS, presented chronologically

Author/Year	Interventions	Sample Size	Outcome Variables	Conclusions
Jung et al. 1980	IS IPPB	126	PPC, CXR	No differences
Morran 1983	Control DBE	102	PPC, CXR, ABG	DBE superior to no treatment
Stock et al. 1984	CPAP IS DBE	63	PPC, CXR, PFT	CPAP superior to both IS and DBEX
Celli 1984	Control IPPB IS DBE	172	PPC, CXR	All treatments superior to no treatment
Hallbrook et al. 1984	Pre-op DBE DBE, cough, PD DBE, cough, PD & bronchodilator	137	ABG, CXR	No differences
Ricksten 1986	IS CPAP PEP	45	[A-a] O_2 diff, FVC	PEP and CPAP superior to IS
Schwieger 1986	Control IS	40	CXR, ABG, PFT	No differences
Roukema 1988	Control IS	153	CXR, ABG	DBE superior to no treatment
Hall et al. 1991	IS DBE	876	PPC, CXR, PaO_2 LOS	No differences
Christensen 1991	DBE DBE & PEP DBE & IR-PEP	365	PPC, PFT, LOS	No differences
Condie et al. 1993	DBE pre-op DBE, pre- and post-op.	330	PPC	No differences
Hall et al. 1996	Low risk group: IS, DBE High risk group: IS, IS & Physio	456	PPC, CXR, ABG	Low risk: DBE and cough superior to IS High risk: IS superior to IS and physio
Chumillas 1998	Control Physio with DBE	81	PPC, CXR, ABG, FVC	Physio superior to no treatment
Olsen 1997	Control PEP or IR-PEP	368	PPC, SpO_2, FVC	Pre-op physio superior to no treatment

(*Continued*)

Table 3.5. (*Continued*)

Author/year	Interventions	Sample size	Outcome variables	Conclusions
Denehy 2001	DBE CPAP 15 min CPAP 30 min	50	PPC, CXR, SpO₂	No difference
Mackay 2005	Early mobilisation Early mobilisation & DBE	56	PPC	No difference

[A-a] 0₂ diff: Alveolar-arterial oxygen difference; ABG: Arterial blood gases; CPAP: Continuous positive airway pressure; CXR: Chest radiograph; DBE: Deep breathing exercises; IPPB: Intermittent positive pressure breathing; IS: Incentive spirometry; IR-PEP: Inspiratory resistance positive expiratory pressure; LOS: Length of post-operative hospital stay; PaO₂: Partial pressure of arterial oxygen; PD: Postural drainage; PEP: Positive expiratory pressure; PFT: Pulmonary function tests, Physio: Physiotherapy; Post-op: Post-operative; Pre-op: Pre-operative; SpO₂: Oxyhaemoglobin saturation.

(National Health and Medical Research Council 1999 **C**), that no particular physiotherapy technique appears to be more effective than the others in preventing PPC. This is supported by a meta-analysis and a systematic review (Thomas & McIntosh 1994 **R**) concluding that there were no significant differences in the incidence of PPC using either IS, DBE or IPPB. Furthermore they found that IS and DBE were both more effective than no treatment. A second, more recent systematic review examined the effect of IS in preventing PPC (Overend et al. 2001 **R**). The authors found that the balance of evidence from the best available studies (10 out of 46 studies) failed to support the use of IS for decreasing the incidence of PPC following UAS. Both these systematic reviews examined the use of incentive spirometry, however state conflicting conclusions.

Comparisons of pre-operative treatment, CPAP, PEP and early mobilisation have not been the subject of a systematic review to date and will be discussed separately below.

PRE-OPERATIVE TREATMENT

Pre-operative instruction alone was as effective as pre- and post-operative physiotherapy in minimising the incidence of PPC in 48 low risk subjects following cholecystectomy (Bourn et al. 1991 **A**). This result has recently been supported in a study of 102 subjects randomly allocated to receive either pre-operative treatment alone or both pre- and post-operative physiotherapy treatment (Denehy 2002 **A/R**). The authors found no significant difference in the incidence of PPC between the groups. The subjects comprised approximately 60 % having colorectal surgery and 30 % hepatobiliary surgery. A benefit of pre-operative management is that it allows assessment of risk factors for developing PPC and allows clinicians to plan and allocate staffing resources. Further research into the treatment benefits of pre-operative physiotherapy alone is warranted.

CPAP

The effects of CPAP application on lung volumes are well documented in the literature. These include an increase in VC (Lindner et al. 1987 **A**), a reduction in respiratory rate (Putensen et al. 1993 **A**), reduced minute ventilation (MV) (Kesten & Rebuck 1990 **A**) and increased FRC (Andersen et al. 1980 **A**; Lindner et al. 1987 **A**; Putensen et al. 1993 **A**; Stock et al. 1985 **A**). This increase in FRC leads to a reduction in shunt, improved SpO_2 and lung compliance, and a decrease in the work of breathing (Dehaven et al. 1985 **A**; Nunn 1993 **C**; Williamson & Modell 1982 **A**).

The application of CPAP following UAS has been demonstrated to increase FRC when compared with other forms of respiratory prophylaxis (Lindner et al. 1987 **A**; Stock et al. 1985 **A**). However, no benefits were reported by Carlsson and co-workers (1981 **A**) when studying a similar patient population.

There is also support for the improvement of atelectasis with CPAP application after UAS (Andersen et al. 1980 **A**; Duncan et al. 1987 **A**; Stock et al. 1985 **A**; Williamson & Modell 1982 **A**). However, the incidence of PPC does not appear to be influenced by the dosage and frequency of application of CPAP in these comparative studies. CPAP appears to be effective in improving lung volumes more quickly than voluntary inspiratory manoeuvres, but this may not have important clinical ramifications (O'Donohue 1992 **R**). In a study of 50 subjects having UAS, Denehy and colleagues (2001 **A**) found no significant difference in FRC, PPC or LOS between subjects receiving physiotherapy (comprising deep breathing exercises and early mobilisation) and those receiving CPAP. The sample size in this study was small however, and significant results were not obtained.

In a sample of 209 UAS patients admitted to intensive care with acute hypoxaemia, the application of CPAP has been demonstrated to reduce the requirement for intubation and the incidence of severe complications (Squadrone et al. 2005 **A**). The use of CPAP in intensive care units is generally considered to be a medical intervention and therefore physiotherapists may not be involved in the decision making process.

PEP MASK

Physiotherapy treatment with the PEP mask was pioneered in Denmark (Falk et al. 1984 **A**). The research evidence examining the efficacy of PEP mask physiotherapy is conflicting and has primarily been conducted in patients with chronic sputum production. The effect of adding PEP to conventional physiotherapy was assessed in a study of 71 patients following elective UAS (Campbell et al. 1986 **A**). The incidence of PPC was found to be 31 % in the group receiving conventional physiotherapy treatment and 22 % in the group receiving physiotherapy plus PEP treatment. The PEP device used in this study is now commonly known as 'bubble' PEP, with positive expiratory pressure maintained by the height of a column of water in a plastic bottle. In this study, a manometer was not added to the PEP circuit, therefore it is unknown if sufficient positive pressures were maintained during treatment. Several other methodological flaws also limit generalisability of the findings of this study.

In a comparison of PEP mask physiotherapy with CPAP and IS, Ricksten and colleagues (1986 **A**) concluded that PEP and CPAP were significantly more effective than IS in maintaining post-operative gas exchange and lung volumes, and lowering the incidence of atelectasis in 43 patients undergoing elective UAS. The results from this well controlled study suggest that both post-operative CPAP and PEP may be equally effective in PPC prophylaxis. The physiological mechanisms responsible for the effectiveness of PEP are thought to work by lung recruitment through collateral channels, however few studies have investigated the effects of PEP on physiological parameters (Van Hengstrum et al. 1991 **A**).

Olsen and colleagues compared PEP and inspiratory resistance PEP to no treatment and as previously discussed, demonstrated a significantly reduced incidence of PPC in treatment groups (Olsen et al. 1997 **A**). In a second study, the same author compared PEP with CPAP in 70 subjects undergoing thoracoabdominal surgery (Olsen et al. 2002 **A**). The application of CPAP for three days followed by PEP therapy decreased the risk of reintubation when compared to PEP therapy alone.

While the use of PEP in the management of patients undergoing UAS has been supported in these studies, the extent of its use in UAS in clinical practice has not been examined. Comparison of PEP therapy with other more simple techniques such as deep breathing exercises may be warranted.

POSITIONING AND MOBILISATION

Upright positioning and mobilisation are frequently utilised by physiotherapists in the post-operative treatment of impaired ventilation. It has been well established that upright positioning is superior to the supine position in improving pulmonary function. Nielsen and colleagues (2003 **R**) in their systematic review concluded that in the post-operative period, upright positioning significantly improves FRC, SpO_2 and PaO_2, and reduces $PaCO_2$. In the literature, upright positioning encompasses sitting, standing and even ambulation. It is yet to be established whether one of these positions is superior in its effects on post-operative pulmonary function. A trend for an increase in minute ventilation due to augmentation of both tidal volume and respiratory rate when progressing from sitting to standing to marching on the spot has been demonstrated (Orfanos et al. 1999 **A**; Zafiropoulos et al. 2004 **A**). However, it appears these differences are not significant, and direct comparisons involving large samples of patients have not been undertaken.

Early mobilisation is an important and widely practised component of post-operative patient care following UAS. Its benefits were first reported in the 1940s, when early mobilisation was observed to hasten post-operative recovery of strength and morale, reduce pressure on hospital beds and nursing services, and most importantly, reduce the incidence of post-operative pulmonary and vascular complications without adverse effects (Brieger 1983 **R**).

In many hospitals, physiotherapists play a major role in the early instigation of patient mobilisation following UAS. A recent survey of Australian physiotherapists working in major public hospitals found that 92 % of respondents 'always' include mobilisation in their post-operative treatment of UAS patients, with the remainder of respondents including mobilisation 'often' (Browning 2005 **A**). Physiotherapists reported the main reasons for the use of mobilisation to be the optimisation of ventilation and prevention of PPC.

Questions have been raised about the appropriateness of physiotherapy interventions involving mobilisation (Dean & Ross 1992 **R**). It is common practice for physiotherapists to assist patients in mobilising with little use of structured programmes or objective measures. Little attention has been given to the intensity of mobilisation and its effect on pulmonary function. It has been suggested that pulmonary function may be improved through the use of structured mobilisation programmes of sufficient intensity (Orfanos et al. 1999 **A**), but this is yet to be formally investigated.

In a number of randomised trials demonstrating that post-operative physiotherapy is effective in reducing the incidence of PPC, early mobilisation has been included in the physiotherapy treatment regimen (Celli et al. 1984 **A**; Olsen et al. 1997 **A**; Roukema et al. 1988 **A**). It is unknown whether techniques such as deep breathing exercises or the early mobilisation included in these interventions were responsible for the reduction in PPC.

In a recently published trial, the addition of deep breathing exercises to a physiotherapist directed programme of early mobilisation was found to have no additional effect on reducing the incidence of PPC in 52 open abdominal surgery patients (Mackay et al. 2005 **A**). Similar results have been obtained in research involving open heart surgical patients (Brasher et al. 2003 **A**; Jenkins et al. 1990 **A**; Stiller et al. 1994 **A**).

Upright positioning and early mobilisation play an important role in the recovery of pulmonary function and prevention of PPC following UAS. Due to advances in analgesia and post-operative care, mobilisation can be achieved earlier and at a greater intensity and frequency.

As they appear equally efficacious, any of the physiotherapy techniques discussed above may be employed in the physiotherapy management of Mr C. With a physiotherapist already on staff, it may be more cost effective to utilise the therapist's manual skills, rather than purchase equipment or choose complicated techniques that may take longer to implement.

Therefore, in the case of Mr C, deep breathing exercises, upright positioning and early mobilisation were the post-operative interventions administered. Preoperatively, instruction and an assessment of risk factors for the development of PPC were also performed.

QUESTION 4

How should these post-operative physiotherapy techniques be administered?

DEEP BREATHING EXERCISES

In normal lungs, regular large breaths to total lung capacity (TLC) are essential to maintain inflation. A study by Ferris and Pollard (1960 **A**) concluded that five consecutive breaths to TLC are necessary to effectively inflate alveoli. In a study of excised lungs from dogs, it was shown that inflated alveoli collapse after one hour when shallow breaths are taken (Anthonisen 1964 **B**). The aim of a deep breath is to produce a large and sustained increase in transpulmonary pressure which distends the lungs and reinflates collapsed lung units (Duggan & Kavanagh 2005 **R**).

A sustained maximal inspiration (SMI) mimics a sigh or yawn and also aims to increase transpulmonary pressure (Bakow 1977 **R**). Sustained maximal inspirations have been reported to redistribute gas into areas of low lung compliance, thus enhancing lung expansion through interdependence using collateral ventilation pathways (Marshall & Widdicombe 1961 **A**; Menkes 1977 **R**; O'Donohue 1992 **R**; Terry et al. 1978 **A**). It may also allow time for alveoli with slow time constants to fill. The addition of a three second SMI at TLC has been recommended in the literature (Bakow 1977 **R**; Terry et al. 1978 **A**).

If regional ventilation is reduced as a result of secretion plugging, the re-expansion of collapsed alveoli may allow air to move behind the secretions and assist their removal using forced expiration techniques (Menkes and Traystman 1977 **R**; Pryor 1991 **C**).

Based on this research from nearly 40 years ago, the common treatment regimen used for breathing exercises is five deep breaths, with a three second SMI, once every waking hour (Bartlett et al. 1973 **R**; Platell & Hall 1997 **R**).

In the systematic review conducted by Thomas and McIntosh (1994 **R**), the regimen of breathing exercises was found to be reasonably uniform across the reviewed studies. This indicates that the treatment regimen discussed above, which was based upon physiological principles and developed in the 1960s, is still common in clinical practice today. The lack of current research evidence to support the method of implementation of breathing exercises means that this technique may be used sub-maximally by physiotherapists and this in turn may reduce treatment efficacy.

There is a paucity of literature evaluating different methods of applying breathing strategies by physiotherapists. It is unclear whether it is more effective to teach deep breathing exercises by encouraging greater abdominal excursion or facilitating bilateral costal (bucket handle) movement, or whether just asking the patient for a maximal inspiration is sufficient. The results of Blaney and colleague (Blaney & Sawyer 1997 **A**) demonstrate a significant increase in diaphragmatic excursion with a 'hands on' approach to breathing exercises following surgery. In this study, verbal instruction alone was compared with two tactile or 'hands on' breathing techniques.

MOBILISATION

Despite the frequent inclusion of mobilisation as a component of physiotherapy treatment in the literature, an examination of its optimal prescription in the post-operative period is yet to be undertaken. As stated by Dean and Ross (1992 **R**), 'the classic unstructured low intensity hallway ambulation is not considered a potential therapeutic intervention and does not constitute an effective use of the therapists' expertise and time'. Therefore, when mobilisation is administered as a post-operative physiotherapy technique, a structured approach is recommended.

Mackay and colleagues (2005 **A**), in their randomised trial, administered a standardised programme of early mobilisation in 52 subjects following UAS. As part of the programme, subjects were encouraged to achieve one or more progressive mobility goals and to walk at a speed where they were taking deeper breaths with an intensity of at least 6/10 according to the modified Borg scale (Borg 1982 **R**). Interventions were administered three times daily on post-operative days one and two, twice daily on days three and four, and then daily until the patient was independently mobile. As all subjects participating in the study received this standardised intervention, it is not possible to compare the benefits of this programme with other mobilisation regimens.

The patient's capacity for mobilisation in the early post-operative period needs to be considered. Post-operative care pathways, which include high quantities of mobilisation, have been reported in recent surgical literature. These pathways report mobilisation of up to 60 metres five times daily, commencing on the first post-operative day (Delaney et al. 2001 **A**). These studies show that high quantities of mobilisation are possible following UAS, however they conflict with recent research conducted by Browning and colleagues (2006 **A**). In this study, the quantity of upright mobilisation achieved in the early post-operative period following UAS was measured. The sample of 50 subjects from a tertiary Australian hospital achieved median upright mobilisation times of 3.2, 7.6, 13.4 and 34.4 minutes on the first to fourth post-operative days respectively. These values were lower than expected and indicate that low quantities of post-operative upright mobilisation are currently being achieved. Structured pathways or mobilisation programmes did not form part of the post-operative care of the patients examined. A significant finding of this study was that the quantity of time spent upright was found to be a significant predictor of hospital LOS ($p < 0.001$), with patients who were more active likely to require shorter admission times. Therefore, increasing the quantity of mobilisation may have a positive effect on improving significant outcomes such as post-operative LOS.

In the early post-operative period, UAS patients can be considered as acutely unwell, therefore care must be taken with the administration of intensive mobilisation. Guidelines for the safe implementation of mobilisation in acutely unwell patients have been published (Stiller & Phillips 2003 **R**). According to these guidelines, monitoring of physiological responses to mobilisation is necessary.

In the case of Mr C, a structured approach to mobilisation was taken. A mobilisation programme was administered twice daily for the first three post-operative days by the physiotherapist, and thereafter by a physiotherapy assistant. The modified Borg scale was used to describe optimal intensity during mobilisation. Measurements of heart rate, blood pressure, oxygen saturation and pain were taken before and after intervention to assess the patient's physiological response to exercise.

It is recognised that this approach to mobilisation requires a significant time commitment from the therapist. In addition to examining the benefits of a structured intensive mobilisation programme, future research should consider whether this can be competently administered by other health care workers.

QUESTION 5

Is physical function impaired following UAS? How can this be minimised?

With advances in anaesthesia, surgical techniques and perioperative care, together with the aging population, it is becoming more common for elderly patients to successfully undergo major surgery. Not only are these patients at a higher risk of developing PPC (Brooks-Brunn 1995a **R**), but they are also at risk of a significant decline in physical function. A large study examining functional recovery of patients aged over 60 years following UAS found that at six months post-surgery, return to pre-operative levels was not achieved by 39 % of patients in the timed up and go test, 58 % of patients in the functional reach test, and 52 % of patients in a grip strength test (Lawrence et al. 2004 **A**). The mean time for recovery of activities of daily living in this sample was three months, and for recovery of instrumental activities of daily living it was six months. This study confirms that physical disability post major abdominal surgery persists for many months post hospital discharge.

In a study of coronary artery bypass patients comparing a high frequency to a low frequency physiotherapy mobilisation programme, it was found that subjects in the high frequency group achieved functional milestones such as sitting in a chair and walking in the ward earlier (van der Peijl et al. 2004 **A**).

Early mobilisation plays a role in minimising the weakness and debility experienced by many patients following major surgery. In a group of patients undergoing elective colorectal surgery, Henriksen and colleagues (2002 **A**) compared enforced mobilisation, pre-operative education and optimal analgesia (intervention group) to standard care (control group). The intervention group spent a significantly greater proportion of time sitting out of bed and ambulating, and recorded significantly less reduction in knee extension strength at seven days and one month. Two months following surgery, strength was 15 % greater than pre-operative values.

Although early mobilisation forms part of routine post-operative nursing care in many hospitals throughout the world, the addition of intensive mobilisation, which is often instigated by physiotherapists, appears to have a positive effect on the return of physical function following UAS. The extent of this effect, both in the immediate post-operative period and post hospital discharge has yet to be formally investigated.

QUESTION 6

What evidence is there to support prophylactic physiotherapy intervention in other surgical populations?

Following any major surgical procedure, the pathophysiological effects of anaesthesia and the perioperative process are similar. For cardiac, thoracic and oesophageal surgery, different factors may influence post-operative outcomes when compared with UAS, but to date only the role of physiotherapy in cardiac surgery has been studied extensively.

From the evidence presented it is clear that some form of physiotherapy intervention is necessary following UAS to prevent PPC occurring in approximately one of every four to five patients treated. However, following cardiac surgery, the evidence obtained from a large body of research clearly challenges the continued traditional necessity for prophylactic physiotherapy intervention. Stiller and colleagues (1994 **A**) found no difference in the incidence of PPC between a group of cardiac surgery patients receiving physiotherapy and a group receiving no physiotherapy intervention. A recent systematic review concluded that there is no clear evidence that prophylactic respiratory physiotherapy reduces the incidence of PPC following cardiac surgery (Pasquina et al. 2003 **A**).

Physiotherapy for thoracic surgical patients has been strongly advocated in several studies, yet little supporting evidence in the form of randomised clinical trials exists. There is no study that uses a no treatment control in the thoracic surgery literature. This no doubt reflects the premise that these patients are at high risk of developing PPC and therefore it would be unethical to withhold treatment. One study evaluated the efficacy of post-operative physiotherapy using IS compared with DBE (Gosselink et al. 2000 **A**). No significant difference in the incidence of PPC between the two treatment groups was demonstrated. The incidence of PPC was 8 % in the 40 subjects who underwent thoracic surgery. These results are in contrast to those of Wang and co-workers (1999 **A**) who found the incidence of PPC was 32.5 % in subjects undergoing lung resection. This study did not mention if any form of post-operative physiotherapy was instituted.

In the study by Gosselink and colleagues (2000 **A**), a subgroup of 27 subjects having transthoracic resection of the oesophagus had a PPC incidence of 19 % suggesting that this group is at higher risk of developing PPC. Ingwersen and colleagues (1993 **A**) compared the use of CPAP, PEP and inspiratory resistance PEP in a subgroup of 59 subjects having pulmonary resection and found no significant differences between the three treatment interventions in the incidence of PPC.

The literature pertaining to the role of physiotherapy for patients having thoracic and oesophageal surgery is inconclusive. A randomised clinical trial examining physiotherapy intervention is much needed in these patient populations.

QUESTION 7

What do we expect to see in surgery in the future? How will this affect physiotherapists?

Recent surgical literature has focussed on the implementation of fast track or multi-modal clinical pathways. These pathways aim to accelerate recovery, reduce morbidity and significantly decrease hospital LOS through the use of optimal pain relief, regional anaesthesia, minimally invasive surgery, early enteral nutrition and enforced early mobilisation (Fearon et al. 2005 **R**). Early mobilisation described in these pathways typically involves commencement of mobilisation on the day of surgery, and encouragement of patients to commence ambulating one circuit of the ward up to five times on the first post-operative day and to sit out of bed for as long as 12 hours daily (Delaney et al. 2001 **A**). Reductions in LOS to as little as two days have been demonstrated with multi-modal clinical pathways post open and laparoscopic colorectal surgery (Andersen & Kehlet 2005 **A**; Bardram et al. 1995 **A**; Basse et al. 2004 **A**; Kehlet & Mogensen 1999 **A**; Moiniche et al. 1994 **A**). Unfortunately, physiotherapy has little mention in this literature to date and it is unclear if any form of physiotherapy intervention is administered.

In many hospitals throughout the world, surgery that was previously performed via a large incision is now more commonly performed laparoscopically. It has been well established in the literature that laparoscopic cholecystectomy is associated with a low incidence of PPC (Hall et al. 1996a **A**), and in Australia routine physiotherapy intervention is not administered to this patient group. A narrative literature review conducted by Olsen et al. (1999 **A**) concluded that routine prophylactic chest physiotherapy is not necessary after laparoscopic upper gastro-intestinal surgery such as fundoplication and vertical banded gastroplasty. The efficacy of physiotherapy in other forms of laparoscopic surgery such as colorectal surgery has not been investigated. A recent survey found that 58 % of physiotherapists in Australian hospitals where laparoscopic colorectal surgery is performed routinely assess and treat these patients post-operatively (Browning 2005 **A**). Future research examining the need for physiotherapy in this patient group is recommended.

A priority for the health care system is the reduction of waiting list times for elective surgery. A proposed method of reducing morbidity pre- and post-surgery, and accelerating post-operative recovery is the use of progressive exercise prehabilitation programmes (Carli & Zavorsky 2005 **R**). Although prehabilitation is common in orthopaedic surgery, similar programmes have not yet been investigated in the abdominal surgery population.

CONCLUSION

With shortages in physiotherapy throughout Australia and in many other parts of the world, the need for routine physiotherapy intervention for patients both before and after major surgery will become increasingly questioned. The use of outcome measures to justify the benefits of physiotherapy interventions will be increasingly more popular. Alternatively, roles which do not require a qualified physiotherapist to perform them may be assigned to other health care workers.

There are many opportunities for research examining the role of the physiotherapist in patients undergoing abdominal surgery and it is certain that the outcomes of such research together with the technological advances in surgery and pain management will define our practice in the future.

REFERENCES

Alexander GD, Schreiner RJ, Smiler BJ (1981) Maximal inspiratory volume and postoperative pulmonary complications. *Surgery, Gynaecology and Obstetrics* **152**: 601–603.

Ali J, Weisel RD, Layug AB, Kripke BJ, Hechtman HB (1974) Consequences of postoperative alterations in respiratory mechanics. *American Journal of Surgery* **128**: 376–382.

American Society of Anaesthesiologists (1963) New classification of physical status. *Anesthesiology* **24**: 111.

Andersen J, Kehlet H (2005) Fast track open ileo-colic resections for Crohn's disease. *Colorectal Disease* **7**: 394–397.

Andersen J, Olesen B, Eikhard B, Jansen E, Qvist J (1980) Periodic continuous positive airway pressure, CPAP, by mask in the treatment of atelectasis. *European Journal of Respiratory Disease* **61**: 20–25.

Anthonisen N (1964) Effect of volume and volume history of the lungs on pulmonary shunt flow. *American Journal of Physiology* **207**: 239.

Australian Institute of Health and Welfare (AIHW) (2005) Australian Hospital Statistics 2003–04. (Health Services Series no. 23) http://www.aihw.gov.au/publications/hse/ahs03-04/ahs03-04-c00.pdf Accessed 1 July 2005.

Bakow ED (1977) Sustained maximal inspiration – a rationale for its use. *Respiratory Care* **22**: 379–382.

Ballantyne J, Carr D, deFerranti S, Suarez T, Lau J, Chalmers T et al. (1998) The comparative effects of postoperative analgesic therapies on pulmonary outcome: Cumulative meta-analyses of randomized controlled trials. *Anesthesia & Analgesia* **86**: 598–612.

Bardram L, Funch-Jensen P, Jensen P, Crawford ME, Kehlet H (1995) Recovery after laparoscopic colonic surgery with epidural analgesia, and early oral nutrition and mobilisation. *Lancet* **345**: 763–764.

Bartlett R (1982) Postoperative pulmonary prophylaxis. *Chest* **81**: 1–3.

Bartlett R, Gazzaniga A, Geraghty T (1973) Respiratory manoeuvres to prevent postoperative pulmonary complications: a critical review. *Journal of the American Medical Association* **224**: 1017–1020.

Basse L, Raskov HH, Jakobsen DH, Sonne E, Billesbolle P, Hendel HW et al. (2002) Accelerated postoperative recovery programme after colonic resection improves physical performance, pulmonary function and body composition. *British Journal of Surgery* **89**: 446–453.

Basse L, Thorbol JE, Lossl K, Kehlet H (2004) Colonic surgery with accelerated rehabilitation or conventional care. *Diseases of the Colon & Rectum* **47**: 271–277.

Blaney F, Sawyer T (1997) Sonographic measurement of diaphragmatic motion after upper abdominal surgery: a comparison of three breathing techniques. *Physiotherapy Theory and Practice* **13**: 207–215.

Borg G (1982) Psychological bases of perceived exertion. *Medicine and Science in Sports and Exercise* **14**: 377–381.

Bourn J, Jenkins S (1992) Post-operative respiratory physiotherapy. *Physiotherapy* **78**: 80–85.

Bourn JEA, Conway JH, Holgate ST (1991) The effect of post-operative physiotherapy on pulmonary complications and lung function after upper abdominal surgery. *European Respiratory Journal* **4**: 325s.

Brasher PA, McClelland KH, Denehy L, Story I (2003) Does removal of deep breathing exercises from a physiotherapy program including pre-operative education and early mobilisation after cardiac surgery alter patient outcomes? *Australian Journal of Physiotherapy* **49**: 165–173.

Brieger GH (1983) Early ambulation. A study in the history of surgery. *Annals of Surgery* **197**: 443–449.

Brooks-Brunn J (1995a) Postoperative atelectasis and pneumonia: risk factors. *American Journal of Critical Care* **4**: 340–349.

Brooks-Brunn JA (1995b) Postoperative atelectasis and pneumonia. *Heart & Lung* **24**: 94–115.

Brooks-Brunn JA (1997) Predictors of postoperative pulmonary complications following abdominal surgery. *Chest* **111**: 564–571.

Browning L (2005) A survey of current mobilisation practices following open upper abdominal surgery. Unpublished.

Browning L, Denehy L, Scholes R (2006) Quantitative measurement of mobility following upper abdominal surgery. *Australian Journal of Physiotherapy* **52**: S8.

Campbell T, Ferguson N, McKinlay R (1986) The use of a simple self-administered method of positive expiratory pressure (PEP) in chest physiotherapy after abdominal surgery. *Physiotherapy* **72**: 498–500.

Carli F, Zavorsky G (2005) Optimizing functional exercise capacity in the elderly surgical population. *Current Opinion in Clinical Nutrition and Metabolic Care* **8**: 23–32.

Carlsson C, Sonden B, Tyhlen U (1981) Can continuous positive airways pressure prevent pulmonary complications after abdominal surgery? *Intensive Care Medicine* **7**: 225–229.

Celli B, Rodriguez K, Snider G (1984) A controlled trial of intermittent positive pressure breathing, incentive spirometry, and deep breathing exercises in preventing pulmonary complications after abdominal surgery. *American Review of Respiratory Disease* **130**: 12–15.

Centre for Evidence-Based Physiotherapy (2006) Physiotherapy Evidence Database (PEDro). http://www.pedro.fhs.usyd.edu.au/index.html Accessed 29 March 2006.

Chumillas S, Ponce JL, Delgado F, Viciano V, Mateu M (1998) Prevention of postoperative pulmonary complications through respiratory rehabilitation: a controlled clinical study. *Archives of Physical and Medical Rehabilitation* **79**: 5–9.

Chuter TA, Weissman C, Starker PM, Gump FE (1988) Diaphragmatic function after cholecystectomy: effect of incentive spirometry. *Current Surgery* **45**: 390–392.

Condie E, Hack K, Ross A (1993) An investigation of the value of routine provision of postoperative chest physiotherapy in non-smoking patients undergoing elective abdominal surgery. *Physiotherapy* **79**: 547–552.

Craig DB (1981) Postoperative recovery of pulmonary function. *Anesthesia and Analgesia* **60**: 46–52.

Dean E, Ross J (1992) Discordance between cardiopulmonary physiology and physical therapy: towards a rational basis for practice. *Chest* **101**: 1694–1698.

Dehaven C, Hurst J, Branson R (1985) Postextubation hypoxemia treated with a continuous positive airway pressure mask. *Critical Care Medicine* **13**: 46–48.

Delaney C, Fazio V, Senagore A, Robinson B, Halverson AL, Remzi FH (2001) 'Fast track' postoperative management protocol for patients with high co-morbidity undergoing complex abdominal and pelvic colorectal surgery. *British Journal of Surgery* **88**: 1533–1538.

Denehy L (2002) *The Physiotherapy Management of Patients Following Upper Abdominal Surgery* PhD thesis, School of Physiotherapy, The University of Melbourne, Melbourne.

Denehy L, Carroll S, Ntoumenopoulos G, Jenkins S (2001) A randomized controlled trial comparing periodic mask CPAP with physiotherapy after abdominal surgery. *Physiotherapy Research International* **6**: 236–250.

Dilworth JP, White RJ (1992) Postoperative chest infection after upper abdominal surgery: an important problem for smokers. *Respiratory Medicine* **86**: 205–210.

Duggan M, Kavanagh BP (2005) Pulmonary atelectasis – a pathogenic perioperative entity. *Anesthesiology* **102**: 838–854.

Duncan S, Negrin R, Mihm F, Guilleminault C, Raffin T (1987) Nasal continuous positive airway pressure in atelectasis. *Chest* **92**: 621–624.

Durreuil B, Cantineau J, Desmonts J (1987) Effects of upper or lower abdominal surgery on diaphragmatic function. *British Journal of Anaesthesia* **59**: 1230–1235.

Fairshter RD, Williams JH (1987) Pulmonary physiology in the postoperative period. *Critical Care Clinics* **3**: 286–306.

Falk M, Kelstrup M, Andersen J, Kinoshita P, Stovring S, Goth I (1984) Improving the ketchup bottle with positive expiratory pressure (PEP) in cystic fibrosis. *European Journal of Respiratory Disease* **65**: 423–432.

Fearon KCH, Ljungqvist O, Meyenfeldt MV, Revhaug A, Dejong CHC, Lassen K et al. (2005) Enhanced recovery after surgery: a consensus review of clinical care for patients undergoing colonic resection. *Clinical Nutrition* **24**: 466–477.

Ferris B, Pollard D (1960) Effect of deep and quiet breathing on pulmonary compliance in man. *Journal of Clinical Investigation* **39**: 143–149.

Forbes AR (1976) Halothane depresses mucociliary flow in the trachea. *Anesthesiology* **45**: 59–63.

Ford GT, Rosenal TW, Clergue F, Whitlaw WA (1993) Respiratory physiology in upper abdominal surgery. *Clinics in Chest Medicine* **14**: 237–252.

Gamsu G, Singer M, Vincent H, Berry S, Nadel J (1976) Post-operative impairment of mucous transport in the lung. *American Review of Respiratory Disease* **114**: 673–679.

Gosselink R, Schrever K, Cops P, Witvrouwen H, De Leyn P, Troosters T et al. (2000) Incentive spirometry does not enhance recovery after thoracic surgery. *Critical Care Medicine* **28**: 679–683.

Grass JA (2005) Patient-controlled analgesia. *Anesthesia and Analgesia* **101**: S44–S61.

Hall JC, Tarala RA, Hall JL (1996a) A case-control study of postoperative pulmonary complications after laparoscopic and open cholecystectomy. *Journal of Laparoendoscopic Surgery* **6**: 87–92.

Hall JC, Tarala RA, Hall JL, Mander J (1991) A multivariate analysis of the risk of pulmonary complications after laparotomy. *Chest* **99**: 923–927.

Hall JC, Tarala RA, Tapper J, Hall JL (1996b) Prevention of respiratory complications after abdominal surgery: a randomised clinical trial. *British Medical Journal* **312**: 148–152.

Hallbook T, Lindblad B, Lindroth B, Wolff T (1984) Prophylaxis against pulmonary complications in patients undergoing gall bladder surgery. *Annales Chirurgiae et Gynaecologiae* **73**: 55–58.

Henriksen MG, Jensen MB, Hansen HV, Jespersen TW, Hessov I (2002) Enforced mobilization, early oral feeding, and balanced analgesia improve convalescence after colorectal surgery. *Nutrition* **18**: 147–152.

Herbert R (2000) How to estimate treatment effects from reports of clinical trials: II dichotomous outcomes. *Australian Journal of Physiotherapy* **46**: 309–313.

Herbert R, Jamtvedt G, Mead J, Birger Hagen K (2005) *Practical Evidence-Based Physiotherapy* London: Elsevier.

Ingwersen UM, Larsen R, Bertelsen MT, Kiil-Nielsen K, Laub M, Sandermann J et al. (1993) Three different mask physiotherapy regimens for prevention of post-operative pulmonary complications after heart and pulmonary surgery. *Intensive Care Medicine* **19**: 294–298.

Jenkins S, Soutar S, Loukota J, Johnson L, Moxhham J (1990) A comparison of breathing exercises, incentive spirometry and mobilisation after coronary artery surgery. *Physiotherapy Theory and Practice* **6**: 117–126.

Jenkins SC, Soutar SA, Loukota JM, Johnson LC, Moxham J (1989) Physiotherapy after coronary artery surgery: are breathing exercises necessary? *Thorax* **44**: 634–639.

Johnson N, Pierson D (1986) The spectrum of pulmonary atelectasis: pathophysiology, diagnosis and therapy. *Respiratory Care* **31**: 1107–1120.

Katagiri H, Katagiri M, Kieser T, Easton P (1998) Diaphragm function during sighs in awake dogs after laparotomy. *American Journal of Respiratory and Critical Care Medicine* **157**: 1085–1092.

Kehlet H (1997) Multimodal approach to control postoperative pathophysiology and rehabilitation. *British Journal of Anaesthesia* **78**: 606–617.

Kehlet H, Mogensen T (1999) Hospital stay of 2 days after open sigmoidectomy with a multimodal rehabilitation programme. *British Journal of Surgery* **86**: 227–230.

Kehlet H, Wilmore DW (2002) Multimodal strategies to improve surgical outcome. *American Journal of Surgery* **183**: 630–641.

Kesten S, Rebuck A (1990) Ventilatory effects of nasal continuous positive airway pressure. *European Respiratory Journal* **3**: 498–501.

Konrad F, Schiener R, Marx T, Georgieff M (1995) Ultrastructure and mucociliary transport of bronchial respiratory epithelium in intubated patients.*Intensive Care Medicine* **21**: 482–489.

Lansing A, Jamieson W (1963) Mechanisms of fever in pulmonary atelectasis. *Archives of Surgery* **87**: 184–190.

Lawrence VA, Hazuda HP, Cornell JE, Pedersen T, Bradshaw P, Mulrow CD et al. (2004) Functional independence after major abdominal surgery in the elderly. *Journal of the American College of Surgeons* **199**: 762–772.

Lawrence VA, Smetana GW, Cornell JE (2005) Prevention of post-operative pulmonary complications: a systematic review. *Journal of General Internal Medicine* **20**: 87–88.

Lindner K, Lotz P, Ahnefeld F (1987) Continuous positive airway pressure effect on functional residual capacity, vital capacity and its subdivisions. *Chest* **92**: 66–70.

Mackay M, Ellis E, Johnston C (2005) Randomised clinical trial of physiotherapy after open abdominal surgery in high risk patients. *Australian Journal of Physiotherapy* **51**: 151–159.

Mackay MR, Ellis E (2002) Physiotherapy outcomes and staffing resources in open abdominal surgery patients. *Physiotherapy Theory and Practice* **18**: 75–93.

Marini JJ (1984) Postoperative atelectasis: pathophysiology, clinical importance, and principles of management. *Respiratory Care* **29**: 516–528.

Marshall R, Widdicombe J (1961) Stress relaxation of the human lung. *Clinical Science* **20**: 19–31.

McAlister F, Bertsch K, Man J, Bradley J, Jacka M (2005) Incidence of and risk factors for pulmonary complications after nonthoracic surgery. *American Journal of Respiratory and Critical Care Medicine* **171**: 514–517.

Menkes H, Traystman J (1977) Collateral Ventilation. *American Review of Respiratory Disease* **116**: 287–309.

Meyers J, Lembeck L, O'Kane H, Baue A (1975) Changes in functional residual capacity of the lung after operation. *Archives of Surgery* **110**: 576–582.

Moiniche S, Dahl JB, Rosenberg J, Kehlet H (1994) Colonic resection with early discharge after combined subarachnoid-epidural analgesia, preoperative glucocorticoids, and early postoperative mobilization and feeding in a pulmonary high-risk patient. *Regional Anesthesia* **19**: 352–356.

Morran CG, Finlay IG, Mathieson M, McKay AJ, Wilson N, McArdle CS (1983) Randomized controlled trial of physiotherapy for postoperative pulmonary complications. *British Journal of Anaesthesia* **55**: 1113–1116.

National Health and Medical Research Council (1999) A guide to the development, implementation and evaluation of clinical practice guidelines. Commonwealth of Australia, Canberra.

Nielsen KG, Holte K, Kehlet H (2003) Effects of posture on postoperative pulmonary function. *Acta Anaesthesiologica Scandinavica* **47**: 1270–1275.

Nunn J (1990) Effects of anaesthesia on respiration. *British Journal of Anaesthesia* **65**: 54–62.

Nunn J (1993) *Nunn's Applied Respiratory Physiology* Oxford: Butterworth-Heineman.

O'Donohue W (1992) Postoperative pulmonary complications. *Postgraduate Medicine* **91**: 167–175.

O'Donohue WJ (1985) Prevention and treatment of postoperative atelectasis: can it and will it be adequately studied? *Chest* **87**: 1–2.

Olsen MF (2000) Chest physiotherapy in open and laparoscopic abdominal surgery. *Physical Therapy Reviews* **5**: 125–130.

Olsen MF, Hahn I, Nordgren S, Lönroth H, Lundholm K (1997) Randomized controlled trial of prophylactic chest physiotherapy in major abdominal surgery. *British Journal of Surgery* **84**: 1535–1538.

Olsen MF, Josefson K, Lonroth H (1999) Chest physiotherapy does not improve the outcome in laparoscopic fundoplication and vertical-banded gastroplasty. *Surgical Endoscopy* **13**: 260–263.

Olsen MF, Wennberg E, Johnsson E, Josefson K, Lonroth H, Lundell L (2002) Randomized clinical study of the prevention of pulmonary complications after thoracoabdominal resection by two different breathing techniques. *British Journal of Surgery* **89**: 1228–1234.

Orfanos P, Ellis E, Johnston C (1999) Effects of deep breathing exercises and ambulation on pattern of ventilation in post-operative patients. *Australian Journal of Physiotherapy* **45**: 173–182.

Overend TJ, Anderson CM, Lucy SD, Bhatia C, Jonsson BI, Timmermans C (2001) The effect of incentive spirometry on postoperative pulmonary complications: a systematic review. *Chest* **120**: 971–8.

Pasquina P, Tramer MR, Walder B (2003) Prophylactic respiratory physiotherapy after cardiac surgery: systematic review. *British Medical Journal* **327**: 1379–1381.

Pasteur W (1910) Active lobar collapse of the lung after abdominal surgery. *Lancet* **2**: 1080–1083.

Platell C, Hall JC (1997) Atelectasis after abdominal surgery. *Journal of the American College of Surgeons* **185**: 584–592.

Pryor J (1991) In: Pryor J (ed.) *Respiratory Care Vol. 7* London: Churchill Livingstone, pp. 79–99.

Putensen C, Hormann C, Baum M, Lingnau W (1993) Comparison of mask and nasal continuous positive airway pressure after extubation and mechanical ventilation. *Critical Care Medicine* **21**: 357–363.

Rankin SL, Briffa TG, Morton AR, Hung J (1996) A specific activity questionnaire to measure the functional capacity of cardiac patients. *American Journal of Cardiology* **77**: 1220–1223.

Richardson J, Sabanathan S (1997) Prevention of respiratory complications after abdominal surgery. *Thorax* **52**: S35–S40.

Ricksten S, Bengtsson A, Soderberg C, Thorden M, Kvist H (1986) Effects of periodic positive airway pressure by mask on postoperative pulmonary function. *Chest* **89**: 774–781.

Ridley S (1998) In: Pryor J, Webber B (eds) *Physiotherapy for Respiratory and Cardiac Problems* London: Churchill Livingstone, pp. 295–327.

Roukema J, Carol E, Prins J (1988) The prevention of pulmonary complications after upper abdominal surgery in patients with noncompromised pulmonary status. *Archives of Surgery* **123**: 30–34.

Sackett D, Strauss S, Richardson W, Haynes R (2000) *Evidence-Based Medicine* Edinburgh: Churchill-Livingston.

Scholes R (2005) *Pulmonary Risk Prediction in the Upper Abdominal Surgery Population* PhD thesis, School of Physiotherapy, The University of Melbourne, Melbourne.

Scholes R, Denehy DL, Sztendur E, Browning L (2006) Development of a risk assessment model to predict pulmonary risk following upper abdominal surgery. *Australian Journal of Physiotherapy* **52**: S26.

Shea RA, Brooks JA, Dayhoff NE, Keck J (2002) Pain intensity and postoperative pulmonary complications among the elderly after abdominal surgery. *Heart & Lung* **31**: 440–449.

Simmoneau G, Vivien A, Sartene R, Kunstlinger F, Samii K, Noviant Y, Duroux P (1983) Diaphragm dysfunction induced by upper abdominal surgery. *American Review of Respiratory Disease* **128**: 889–903.

Smith M, Ellis E (2000) Is retained mucus a risk factor for the development of post-operative atelectasis and pneumonia? Implications for the physiotherapist. *Physiotherapy Theory and Practice* **16**: 69–80.

Squadrone V, Coha M, Cerutti E, Schellino MM, Biolino P, Occella P et al. (2005) Continuous positive airway pressure for treatment of postoperative hypoxemia: a randomized controlled trial. *JAMA* **293**: 589–595.

Stiller K, Crawford R, McInnes M, Montarello J, Hall B (1995) The incidence of pulmonary complications in patients not receiving prophylactic chest physiotherapy after cardiac surgery. *Physiotherapy Theory and Practice* **11**: 205–208.

Stiller K, Montarello J, Wallace M, Daff M, Grant R, Jenkins S et al. (1994) Efficacy of breathing and coughing exercises in the prevention of pulmonary complications after coronary artery surgery. *Chest* **10**: 741–747.

Stiller K, Munday R (1992) Chest physiotherapy for the surgical patient. *British Journal of Surgery* **79**: 745–749.

Stiller K, Phillips A (2003) Safety aspects of mobilising acutely ill patients. *Physiotherapy Theory and Practice* **19**: 239–257.

Stock M, Downs J, Gauer P, Alster J, Imrey P (1985) Prevention of postoperative pulmonary complications with CPAP, incentive spirometry, and conservative therapy. *Chest* **87**: 151–157.

Terry P, Traystman R, Newball H, Batra G, Menkes H (1978) Collateral ventilation in man. *New England Journal of Medicine* **298**: 10–15.

Thomas JA, McIntosh JM (1994) Are incentive spirometry, intermittent positive pressure breathing, and deep breathing exercises effective in the prevention of postoperative pulmonary complications after upper abdominal surgery? A systematic review and meta-analysis. *Physical Therapy* **74**: 3–16.

Tisi GM (1979) Preoperative evaluation of pulmonary function. *American Review of Respiratory Disease* **119**: 293–310.

Torrington KG, Henderson C (1988) Perioperative respiratory therapy (PORT): a program of preoperative risk assessment and individualised postoperative care. *Chest* **93**: 946–951.

Tsui S, Lee D, Ng K, Chan T, Chan W, Lo J (1997) Epidural infusion of bupivacaine 0.0625 % plus fentanyl 3.3 micrograms/ml provides better postoperative analgesia than patient-controlled analgesia with intravenous morphine after gynaecological laparotomy. *Anaesthesia and Intensive Care* **25**: 476–481.

Van De Water JM (1972) Preoperative and postoperative techniques in the prevention of pulmonary complications. *Symposium on Respiratory Care in Surgery* 1339–1348.

van der Peijl ID, Vlieland TPM, Versteegh MIM, Lok JJ, Munneke M, Dion RAE (2004) Exercise therapy after coronary artery bypass graft surgery: a randomized comparison of a high and low frequency exercise therapy program. *Annals of Thoracic Surgery* **77**: 1535–1541.

Van Hengstrum M, Festen J, Beurskens C, Hankel M, Beekman F, Corstens F (1991) Effect of PEP versus forced expiration technique on regional lung clearance in chronic bronchitis. *European Respiratory Journal* **4**: 651–654.

Wahba R (1991) Perioperative functional residual capacity. *Canadian Journal of Anaesthesia* **38**: 384–400.

Wang J, Olak J, Ultmann R, Ferguson M (1999) Assessment of pulmonary complications after lung resection. *Annals of Thoracic Surgery* **67**: 1444–1447.

Werawatganon T, Charuluxanun S (2005) Patient controlled intravenous opioid analgesia versus continuous epidural analgesia for pain after intra-abdominal surgery. *Cochrane Library* http://www.thecochranelibrary.com.

Williamson D, Modell J (1982) Intermittent continuous positive airway pressure by mask. *Archives of Surgery* **117**: 970–972.

Wilson R (1983) Intermittent CPAP to prevent atelectasis in postoperative patients. *Respiratory Care* **28**: 71–73.

Zafiropoulos B, Alison JA, McCarren B (2004) Physiological responses to the early mobilisation of the intubated, ventilated abdominal surgery patient. *Australian Journal of Physiotherapy* **50**: 95–100.

III Neurological

4 Practice and Feedback for Training Reach-to-Grasp in a Patient with Stroke

PAULETTE M. VAN VLIET AND KATHERINE DURHAM

CASE REPORT

BACKGROUND

Mrs PJ was a 67 year old woman who lived with her husband in a two-storey house. At the time of the stroke, she was independent in self-care. Her husband was well and had retired from work. She had a daughter who was a regular visitor, lived nearby and was willing to assist in her mother's rehabilitation.

MAIN DIAGNOSIS

A CAT scan within the first few weeks after the stroke revealed 'a wedge shaped low attenuation in the right parietal lobe, consistent with an infarct. There was a focal area of high attenuation in the right basal ganglia with a little low attenuation just anterior to this, which could indicate a small intracerebral haemorrhage, without significant midline shift or mass effect. There were cerebral atrophic changes consistent with the patient's age.'

PREVIOUS MEDICAL HISTORY

Prior to the stroke, the patient had angina, hypertension, coronary artery bypass graft and chronic obstructive airways disease. No previous stroke had occurred.

PRESENTING SYMPTOMS ON ADMISSION

Mrs PJ was admitted to hospital with dysarthria and weakness in her left upper and lower limbs. She also had a left facial weakness, dysphagia and decreased sensation. There was no unilateral spatial neglect or dysphasia and her visual fields were normal.

Recent Advances in Physiotherapy. Edited by C. Partridge
© 2007 John Wiley & Sons, Ltd

Table 4.1. Assessment results

Assessment Tool	Result
Rivermead Motor Assessment (arm section) (Lincoln & Leadbitter 1979)	Cumulative score of 8 (highest level of performance 'Pick up a piece of paper from table in front and release five times').
Modified Ashworth scale (Bohannon & Smith 1987)	Wrist flexors = 1, Finger flexors = 0, Elbow flexors = 1 (1 = 'Slight increase in tone, manifested by a catch and release or by minimal resistance at the end of range of motion when the affected part is moved in flexion or extension', 0 = 'no increase in tone').
Short-Form McGill Pain Questionnaire (Melzack 1987)	2, i.e. discomforting, describing pain on lateral upper arm when performing shoulder forward flexion.
Extended Activities of Daily Living scale (Nouri & Lincoln 1987)	Mobility = 7/18, Kitchen = 11/15, Domestic = 4/15, Household = 6/18. Particular upper limb activities which the patient was unable to complete on this scale included washing up, making a hot snack, doing the housework, using the affected arm to feed herself, and writing.
Rey figure copy for spatial perception (Rey 1959)	26 out of maximum 36.
Star cancellation for neglect (Wilson et al. 1985)	50 out of maximum of 54.
Nottingham Sensory Assessment (Jackson & Crow 1991)	Tactile sensation fingers and hand: light touch = normal, pressure = normal, Kineaesthesis = normal, Two-point discrimination fingertips = impaired (2 points detected but at distance >3 mm).

ASSESSMENT SIX MONTHS AFTER THE STROKE

At the time of the case report, six months had elapsed since the stroke. Mrs PJ remained in hospital for four months. She was now receiving out-patient physiotherapy treatment for her arm twice a week. The assessments shown in Table 4.1 were performed.

ANALYSIS

Detailed assessment of upper limb activities

After joint goal setting with the patient (Blair 1995 **A**; Blair et al. 1996 **A**), it was decided to assess in detail two functional movements of the left arm: 1 reaching for an object in front, and 2 using a fork. Mrs PJ's dominant hand was her left but she also used a fork in the left hand. The analysis involved:

(a) analysis of invariant kinematic features of the movement (compared to normal performance).
(b) identification of kinematic deviations from normal.

(c) tests of the performance of individual components of the movement (including ability to elicit the correct movement, and endurance and strength capacity).

(d) tests of length of individual muscles.

(e) tests of joint mobility.

(f) investigation of other contributing factors such as pain.

(g) identification of the main problem(s) preventing normal performance of the movement.

The analysis has its roots in the seminal work of Carr and Shepherd (2003a **C**). Visual observation was used for (a) and (b) above. The findings of the assessment are summarised in Tables 4.2 and 4.3.

In addition, there were problems with timing. The opening of the hand was delayed relative to the beginning of the transport of the hand to the object. These two events normally begin together as part of a coordinated motor schema in which there is coupling of key temporal events in the grasp and transport components of reach-to-grasp (Hoff & Arbib 1993 **B**).

To improve the accuracy of the observations above, videotape analysis (Van Vliet 1988) or a motion analysis system could be used. The following analysis of using a fork compares the patient's performance to how she was accustomed to using a fork prior to the stroke. It should be acknowledged that there are variations in the way a fork can be used and that a fork may not be the usual eating implement for many people.

Clinical reasoning process used in the analysis

Collaborative reasoning with the patient was used to decide on the activities to target in rehabilitation (Higgs & Jones 2000 **C**). Encouraging the patient to share responsibility for their recovery may improve outcomes after stroke (Partridge & Johnston 1989 **A**). During and following a process of cue identification (for example, kinematic features) and cue interpretation (for example, how these relate to kinematic deviations), multiple hypotheses were formed by inductive reasoning, concerning the possible causes of the absent or reduced kinematic features. Deductive reasoning was then used, where hypotheses were tested as described above and the results of these tests were compared to the initial hypotheses via backward reasoning (Higgs & Jones 2000 **C**). The knowledge base used in this process includes knowledge of the biomechanics of reaching and manipulation and also of the cortical control of reaching, from behavioural and neurophysiological studies. Examples of how this knowledge was used are described in the following sections.

Reaching for a cup – example of clinical reasoning

The patient had difficulty elevating the arm sufficiently. Decreased forward flexion was chosen as a main problem to investigate because based on the observation above, more compensatory strategies were caused as a result of this than other decreased kinematic features (see Table 4.1). In terms of muscle force, this is likely to be due to

Table 4.2. Analysis of reaching for an object in front. Numbers in parentheses link deviation to kinematic feature in previous column

Invariant Kinematic Features	Kinematic Deviations	Muscle Length and Joint Problems
1. Decreased forward flexion at the gleno-humeral joint.	• Excessive elevation of the scapula and abduction at the gleno-humeral joint (1, 2).	• Shortened teres major, subscapularis and latissimus dorsi.
2. Decreased protraction and lateral rotation of the scapula.	• Lateral flexion of the trunk to the right (1, 2).	• Shortened rhomboid major and minor.
3. Decreased external rotation at the gleno-humeral joint.	• Forward flexion of the trunk (1, 2, 4).	• Shortened biceps brachii.
4. Decreased elbow extension (decreased by 10°).	• Pronation of the forearm (5, 8).	• Stiffness in glenohumeral joint.
5. Decreased supination.		• Stiffness in carpal bones of wrist.
6. Decreased radial deviation.		• Shortened pronator teres and pronator quadratus.
7. Decreased wrist extension.		• Shortened adductor pollicis.
8. Decreased abduction and rotation of the carpometacarpal (CMC) joint of the thumb.		• Shortened flexor digitorum superficialis and profundus.
9. Decreased extension of digits 3, 4 and 5.		
10. Decreased flexion at the interphalangeal joint of the thumb and index finger.		

Table 4.3. Analysis of using a fork. Numbers in parentheses link deviation to kinematic feature in previous column

Invariant Kinematic Features	Kinematic Deviations	Muscle Length and Joint Problems
1. Decreased abduction and rotation at CMC joint of thumb to pick up fork.	• Picks up fork with 'hook' type of power grasp* (finger flexion without using thumb) (1, 2).	• Shortened adductor pollicus.
2. Decreased conjunct rotation at MCP joint of index finger to pick up fork.	• Uses less affected arm to position fork in left hand (3).	• Stiffness at carpometacarpal joint. • Stiffness in carpal bones of wrist.
3. Decreased ability to turn fork in hand after picking up.	• Does not place index finger on top of fork – holds with hook grasp (4).	• Shortened internal rotator muscles of shoulder (teres minor, infraspinatus).
4. Decreased extension and abduction at metacarpophalangeal (MCP) joint of index finger to place finger on fork.	• Rocks knife from side to side to cut food, rather than moving back and forth (5, 6).	
5. Decreased flexion of digits 3, 4 and 5 (MCP and IP joints) to hold fork in place in hand.	• Excessive wrist flexion and ulnar deviation (a result of pushing into food with a hook grasp on fork).	
6. Decreased 'cupping' of hand (bringing thenar and hypothenar eminences together).	• Excessive internal rotation and abduction of left glenohumeral joint (as above). • Lateral flexion of trunk to the right (as above).	

*As described by Napier (Napier 1956 B)

decreased force generation in the shoulder flexors (especially anterior deltoid as the prime mover (Basmajian 1976 **B**)), decreased force generation in muscles protracting and laterally rotating the scapula (especially serratus anterior and trapezius, which act as a force couple for scapula setting and movement (Mottram 1997 **C**), and/or decreased force generation in the rotator cuff (which forms a force couple with deltoid to maintain the position of the head of humerus in the glenoid cavity (Nordin & Frankel 2001 **B**)) (especially infraspinatus and teres minor, which ensure full range of elevation by external rotation of the humerus).

Each of these components was tested separately and it was found that Mrs PJ had 90° active forward flexion, 25° active external rotation (50° passive) and 50 % active range of protraction compared to the other side (80 % passive range). Pain was a limiting factor for forward flexion and external rotation. Further tests showed that the rhomboid, teres major, subscapularis and latissimus dorsi muscles were tight, and the glenohumeral joint was stiff compared to the other side when accessory joint mobilisations were performed.

Further investigation of the shoulder pain was performed. This included a subjective and objective examination (Hengeveld & Banks 2005 **C**). The objective examination included passive and active range of motion; strength tests for specific muscles (Cole et al. 1988 **C**) (for example, supraspinatus, biceps, teres minor, infraspinatus, subscapularis); accessory movements of the gleno-humeral, acromioclavicular and sternoclavicular joints; palpation for swelling, wasting and tenderness; and specific tests for subacromial impingement (Neer 1972 **B**), instability (subluxation and anterior and posterior stability (Hawkins & Mohtadi 1991 **B**)), labral tears (Mimori et al. 1999 **B**) and adhesive capsulitis. The subjective findings revealed a gradual onset of pain as elevation recovered after the stroke, and no recollection of a particular event that caused the initial onset of the pain. Objective tests provoked pain on active and passive external rotation, flexion and abduction and internal rotation whilst in 60° abduction (maximum active range; 'empty can' test), and found weakness of external rotators compared to internal rotators, positive Neer impingement sign, and restricted passive elevation when the scapula was prevented from moving. There was no joint instability. It was concluded that a major cause of pain derived from subacromial impingement, involving the supraspinatus tendon and possibly the subacromial bursa. The limited passive range of movement when the scapula was stabilised and the time that had elapsed since the stroke without full active range of movement, also suggested adhesive capsulitis.

The following treatment goals were formed for the problem of decreased forward flexion:

- Improve force generation of teres minor and infraspinatus, anterior deltoid, serratus anterior and trapezius.
- Improve the coordination of transport and grasp components at the beginning of the reach.
- Lengthen teres major, subscapularis, latissimus dorsi and rhomboids.
- Reduce shoulder pain.

Although all these aims are important, the initial priority was to regain external rotation, because this was the most limited, and left as it was would prevent the force couple of rotator cuff and deltoid from working efficiently for forward flexion. It was also likely to be connected with the shoulder pain (Joynt 1992 **A**; Kumar et al. 1990 **A**). Achievement of the first and third goals above was expected to reduce shoulder pain. Additional accessory joint mobilisation were to be employed as necessary to reduce pain and stiffness (Hengeveld & Banks 2005 **C**).

By a similar process of inductive and deductive reasoning, it was decided that another main problem to address in reach-to-grasp was the decreased abduction and rotation of the CMC joint of the thumb.

The main problems with using the fork were reduced conjunct rotation of the index finger to grasp the fork, and decreased extension and abduction of the index finger to place it on top of the fork. Henceforth the discussion will concentrate on training for the reach-to-grasp movement, however the upper limb practice schedule would also include practice to improve use of the fork.

As a prerequisite for practising these activities, synergic muscular activity will also normally occur in other parts of the body to enable forward flexion. Preparatory and ongoing adjustments are normally made to stabilise the trunk. The transversus abdominus is activated in anticipation of any movement to increase intra-abdominal pressure, particularly shoulder flexion (Hodges & Richardson 1999 **B**), and therefore the function of this muscle was assessed. The transversus abdominus was isolated (by locating the anterior superior iliac spine, sliding the hand in and down, and then asking the patient to cough), aiming to dissociate it from the internal obliques where possible. This initially was assessed in crook lying and then in sitting. Mrs PJ was able to activate this muscle and dissociate it from internal obliques. If training was necessary, this would begin in crook lying, then progress to more functional positions and tasks. For example, a progression would be to work the core muscles in sitting, with the arms supported on a high table, and work on pelvis dissociation. This requires the trunk to be the stable reference point and to achieve this requires the activation of the abdominal stabilisers. Bilateral dysfunction may be common after CVA and therefore muscles providing core stability on both sides (transversus abdominus, rectus abdominus, external and internal obliques, and erector spinae muscles) (Creswell et al. 1994 **B**) should be assessed. The sternocleidomastoid and cervical extensor muscles at the neck also demonstrate feedforward activation during rapid unilateral and bilateral upper limb flexion to oppose the reactive forces during arm movements and achieve stability for the visual and vestibular systems during movement (Falla et al. 2004 **B**). As far as possible, these mechanisms were assessed.

TRAINING OF REACH-TO-GRASP

The plan for training will now be described. The focus will be on how practice would be structured and how feedback would be delivered to the patient. This is a proposed training schedule, based on available evidence and knowledge of the patient's problems. Meaningful medium-term goals (Van Vliet et al. 1995 **B**) (for example, 1 week) for training, which relate to treatment goals mentioned earlier and are challenging and achievable, would be set. These would be decided upon jointly by the therapist and patient (Blair 1995 **A**; Blair et al. 1996 **A**). The goals would relate to each of the above mentioned movement problems and be expressed in quantitative terms as much as possible to reduce subjectivity, so that any change was clear to both therapist and patient. The achievement of goals would be evaluated by goal attainment scaling (Reid & Chesson 1998 **A**).

PRACTICE

Content of practice

Active participation should be encouraged with Mrs PJ. A study using transcranial magnetic stimulation (TMS) in healthy subjects has shown that after 30 minutes of training wrist flexion and extension, motor performance improved to a greater extent when the training was active than when it was passive (Lotze et al. 2003 **B**). In another study using TMS in patients with stroke, Hummelsheim showed that active contraction of a muscle led to a larger amplitude and shorter latency of electromyographic output than in more passive methods such as tapping on or weight bearing on the affected arm (Hummelsheim et al. 1995 **A**).

Whole practice for discrete tasks such as reaching for a cup is better than part practice, because the action is planned in advance in an open loop manner via a motor programme (Hoff & Arbib 1993 B; Schmidt & Wrisberg 2000 **C**). If only part of the movement is practised, a different motor programme may be utilised, and so transfer of learning to performance of the whole skill may not naturally occur. The transport and grasp components of reaching are temporally linked at the beginning of the movement and at the time of maximum aperture (Castiello et al. 1993b **B**; Gentilucci et al. 1991 **B**), so whole task practice will allow activation of temporally linked central commands for arm and hand. However, after stroke there may be insufficient force generation in muscles, preventing performance of the whole task, so part practice may be necessary. In that case, the therapist needs to follow part practice with whole task practice in the same session to enable transfer of learning. Mrs PJ's practice contains both whole and part practice. An additional reason to include whole practice is that Mrs PJ's ability to store learned 'chunks', which it has been suggested is necessary for efficient sequence processing, may be impaired as her lesion affects areas of the brain involved in chunking (dominant parietal lobe and basal ganglia) (Kennerley et al. 2003 **B**).

The training exercises are task-specific since this has been shown to be effective for stroke patients (Blennerhassett & Dite 2004 **A**; Platz et al. 2001 **A**; Winstein et al. 2004 **A**). The task-specific approach is supported by cortical mapping studies using transcranial magnetic stimulation, which have demonstrated that the functional organisation of somatosensory cortex may change dynamically according to task requirements by switching between pre-existing maps as necessary (Braun et al. 2001 **B**). Cortical maps in the primary motor cortex also differ between people with different levels of skill (Tyc et al. 2005 **B**). Training has been shown to be specific to joint angle (Sale & MacDougall 1981 **B**), body position (Rasch & Morehouse 1957 **B**), and type and speed of contraction (Rutherford 1988 **B**). Where part practice is used, the practise still has some specificity to the task, for example, external rotation is practised with some forward flexion, since these occur together in reach-to-grasp.

Mental practice would be introduced during periods when actual exercise was precluded by pain, fatigue, or illness. It would not be included otherwise because Mrs PJ was able to attempt all the required movements. Although further mental practice might augment exercise time, Mrs PJ was unlikely to comply with additional practice time. Mental practice can elicit cortical activity in the same brain areas as actual performance (Jeannerod 1994 **B**) and has been found to improve arm movement in two controlled studies (Dijkermann et al. 2004 **A**; Page et al. 2005 **A**).

Evidence that both arms are constrained to behave as coordinated units during bilateral performance of the upper limbs (Castiello et al. 1993a **B**; Tuller et al. 1982 **C**) suggests that bilateral simultaneous practice might drive the activity of the hemiplegic arm by employing undamaged parts of the brain. It could be that by coupling the non-affected with the affected limb, the undamaged hemisphere generates a 'template' for action that facilitates the reorganisation of neural networks within the affected hemisphere. If so, this could be useful in the cognitive stage of learning, when the patient is creating a correct internal representation of the activity (van Wijk F, Personal communication). Several studies provide evidence of improvement from bilateral training after stroke (Cunningham et al. 2002 **A**; Mudie & Matyas 1996 **A**; Whitall et al. 2000 **A**) and another has found that hemiparetic patients demonstrate a temporal coupling between the arms when moving simultaneously (Waller et al. 2006 **B**). In some cases, cortical reorganisation has resulted from repetitive bilateral training with rhythmic auditory cueing (Luft et al. 2004 **A**). Therefore a bilateral task has been included.

Attentional focus of practice

Instruction and feedback about a task can either induce an internal focus (IF) or an external focus (EF) of attention. IF feedback is that which directs attention towards the body's movements whereas EF feedback directs the attention to the effects of the movement on the environment (Magill 2003 **C**). Evidence in healthy subjects shows that EF instruction and feedback induces more effective motor learning (Shea & Wulf 1999 **B**; Wulf & Weigelt 1997 **B**; Wulf et al. 1998 **B**; Wulf et al. 1999 **B**; Wulf et al. 2001b **B**; Zachry et al. 2005 **B**). This evidence supports the use of EF for both novice and skilled tasks and has been found in both the laboratory setting and in practical applications. Zachry et al. (2005 **B**) also found EF increased movement economy, whereas Wulf and Weigelt (1997 **B**) found that IF degraded learning.

It is unclear however, whether the results from research in healthy subjects can be transferred into the neurologically impaired. To date the evidence suggests EF instructions are more effective in patients with stroke in reaching tasks (Fasoli et al. 2002 **A**) and EF feedback is more effective in patients with Parkinson's disease where balance was trained (Landers et al. 2005 **A**). Conversely, there is evidence that IF feedback is effective for training postural control following stroke (McNevin &

Wulf 2003a **A**). It is interesting to note however that the previous study (McNevin & Wulf 2002 **A**), which examined a similar task set up, found that EF improved static balance whereas IF compromised learning. In another study which examined a dual task, walking with a tray, in Parkinson's disease, Canning (2005 **A**) showed that directing the attention to the object, the tray, was more beneficial and enhanced motor performance.

Overall the evidence for the use of EF feedback in healthy subjects is compelling. Although McNevin and Wulf (McNevin et al. 2003a **A**) provide contrary evidence that IF is best, it is conceivable that the benefits of each type of attentional focus feedback may depend upon the stage of recovery. As Mrs PJ was in the later stages of her rehabilitation, encouraging an external focus of attention, particularly whilst providing task instructions, was to be recommended.

The evidence from healthy subjects also demonstrates that increasing the distance of external attention focus enhances learning (McNevin et al. 2003 **B**). With Mrs PJ, focus in the reaching tasks would be encouraged either towards the cup or the placement of the cup, whichever was furthest away. It is interesting to note that Wulf, Shea and Park (Wulf et al. 2001a **B**) found that where subjects were given a choice between IF and EF, EF was chosen more frequently, and those who chose EF were more effective in retention tests than those who chose IF.

Specific exercises

The specific exercises are described below. On some of these, a specific number of repetitions will be requested, as in the second exercise, 'drawing arc'. Figures 4.1 to 4.4 illustrate some of the exercises.

Moving cup out (for external rotation)

Start position: sitting, forearm in sagittal plane and resting on edge of table at side, elbow at 90°, Mrs PJ holds a cup.

Method: colourful stickers on the table represent targets to which to move the cup. These are placed at between 30° and 80° from the sagittal plane in the direction of external rotation (initial attempts at 30° (5° more than current range) and gradually increased to 80° – normal range of other arm).

Instruction: 'I would like you to move the cup to the blue sticker.'

Drawing arc (for external rotation with forward flexion)

Start position: standing, flip chart in front, holding marker pen, shoulder in 60° unsupported forward flexion. A parabolic arc is drawn on the paper, from a position of internal rotation to external rotation of the shoulder.

Instruction: 'I would like you to draw five arcs from the cross on the right to the cross on the left' (towards the left, position stickers to encourage maintenance of forward flexion and external rotation in a pain free and achievable range).

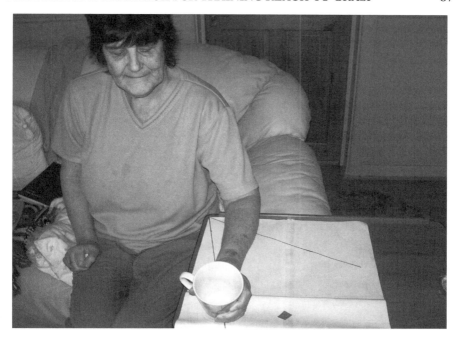

Figure 4.1. Moving cup out exercise.

Figure 4.2. Drawing arc exercise in standing.

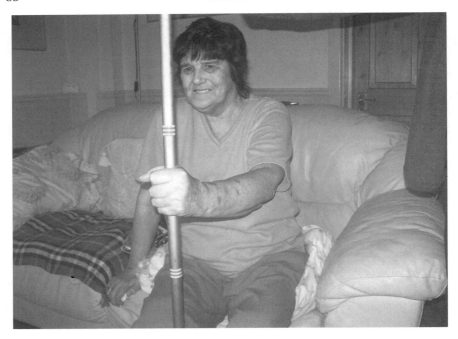

Figure 4.3. Sliding down broom exercise.

Figure 4.4. Moving tray exercise.

Starting off (for coordination of grasp and transport at start of reach)

Start position: sitting, sticker in front of hand on table.
Instruction: 'As you reach towards the cup, ensure your hand is open by the time you pass the sticker.'

Sliding down broom handle (for forward flexion)

Start position: sitting, holding vertical broom handle in front, shoulder initially at 80°.
Method: the hand slides down the handle slowly (eccentric control of anterior deltoid).
   The handle may be gripped when necessary to control the speed of the movement.
Instruction: 'Starting at the first sticker, slide the hand down to the second sticker.'

Getting ready to reach (scapular setting)

Method: all practice tasks need to start from a biomechanically advantageous position. For this I would encourage scapular setting, ensuring the scapular sat on the ribcage and that this position was maintained during reaching. This was achieved by using mirrors both in front of and behind Mrs PJ to show the position of the scapula.

Moving cup forward (protraction)

Start position: sitting, arm resting in front on high table at 90° flexion, elbow extended, holding cup. Target (colourful sticker) is placed to encourage between 50 and 100 % full passive range of protraction, starting at 60 %.
Instruction: 'I would like you to place the cup beyond the sticker.'

Reaching to cup (whole task practice)

Instruction: 'I would like you to reach towards the cup.'

Holding ball (for abduction and conjunct rotation of thumb)

Start position: holding ball 9 cm diameter, with thumb on top of ball.
Method: moves thumb around to side of ball opposite to fingers, with thumb pad in contact with ball at end position.
Instruction: 'I would like you to move your thumb around the ball towards the table, keeping good contact throughout.'

Reaching to can (whole task practice)

Method: wearing a small wrap around splint (made from thermoplastic material) to hold thumb in palmar abduction (Carr & Shepherd 2003a **C**), Mrs PJ reaches to grasp can. The splint is small enough to allow flexion of the interphalangeal joints of thumb and index finger. The can is wide to encourage maximum thumb abduction.
Instruction: 'I would like you to reach to the can.'

Moving tray (bilateral with auditory cueing)

Start position: standing, holding tray on worktop in front with both hands. Husband
    stands in front beyond tray.
Instruction: 'Give the tray to your husband. Then take it back. Keep in time with the
    metronome. Give the tray on the first click. Take it back on the next click.'

Specific stretches

Teres major and subscapularis would be stretched by placing the shoulder in a position
of external rotation, as described by Ada et al. (2005 **A**), where Mrs PJ was supine,
head and shoulders supported with pillows, with the shoulder at 45° abduction and in
external rotation. The maximum passive range was 50°, so a modified position (Ada
et al. used the stretch preventatively, so maximum range was greater than in Mrs PJ)
would be maintained by tying the end of a crepe bandage around the hand and loosely
attaching the other end to the head of the bed, with a pillow under the forearm.
A stretch for latissimus dorsi requires the arm to be held in a position of flexion,
abduction and external rotation. If shoulder pain allowed, the arm would be placed
in this position in supine, with gravity maintaining the stretch. Rhomboid major and
minor would be stretched by positioning the arm in protraction, while resting on a
table at a height of 80° flexion. A marker would indicate where the hand should be
if the shoulder was in maximum protraction, and the patient would note if the hand
moved and either correct the position herself or alert the therapist or assistant. An
air splint might be needed to keep the elbow straight. Adductor pollicus would be
stretched with the use of the small wrap around splint mentioned above.
    Ada et al. (2005 **A**) have shown that 30 minutes of stretch in external rotation, five
days a week for four weeks was sufficient to reduce the development of contractures
in upper limbs which did not yet show signs of contracture. The time would need to
be increased for Mrs PJ, as she already had considerable loss of range of movement.
Forty-five minutes was the maximum time that could be practically managed, so the
external rotation stretch would be maintained for this length of time. The stretch for
latissimus dorsi would be maintained for 30 minutes, because longer might lead to
shoulder discomfort. The two stretches in supine would be carried out at separate
times of the day. In the case of adductor pollicus, the stretching time could occur
during the practice exercises and during the rhomboid stretch. Several short 20 second
stretches would also be given to the internal rotators by the therapist, prior to practice
of external rotation, to decrease stiffness of the muscle (Vattanasilp et al. 2000 **A**).
Mobilisations would also be used to reduce stiffness of the carpal bones of the wrist.
Mrs PJ's husband would be shown how to set up the stretches at home and how to do
the 20 second stretches to internal rotators.

Scheduling of practice

Because the tasks being practised were discrete rather than continuous, learning
was not expected to be adversely affected by fatigue, so the practice could be

massed (practice time > rest time) rather than distributed (rest time > practice time) (Schmidt & Wrisberg 2000 **C**). The one-to-one practice session with the therapist would therefore be carried out in one continuous session, unless the patient's fitness levels or shoulder pain prevented this.

Practice aims to cause not just a change in performance (observable behaviour) (Magill 2007 **C**), but to bring about learning (a relatively permanent improvement in performance) (Magill 2007 **C**). The amount of practice to achieve learning is uncertain but it is probable that many hundreds of repetitions are necessary. Studies of exercise for the upper limb after stroke that have demonstrated a positive outcome show that a minimum of between 30 and 40 minutes of practice of functional tasks (Feys et al. 2004 **A**; Platz et al. 2001 **A**; Sunderland et al. 1992 **A**; Winstein et al. 2004 **A**) or strengthening exercises (Butefisch et al. 1995 **A**; Hummelsheim et al. 1996 **A**; Hummelsheim et al. 1997 **A**) per day for several weeks can be sufficient to make significant gains compared to control subjects. Given that the patient was attending out-patient therapy, the one-to-one sessions with the physiotherapist were likely to be two 45 minute long sessions. These sessions would be extended by attendance at an upper limb practice group for 30 minutes on one day of attendance. Practising with at least one other person, as in a group, can be motivating, increase the feeling of responsibility, and encourage the setting of harder goals, and in healthy people has been shown to be better for learning than practising alone (McNevin et al. 2000 **B**). A semi-supervised stretch session for shortened muscles would also occur on each day of attendance. In semi-supervision, the physiotherapist sets up the position then works with another patient, coming back to check intermittently on the position. The amount of practice would also be increased by self-directed practice at home for 30 minutes on each day of the week. Mrs PJ's husband or daughter would assist with the setting up of each exercise.

To make the home practice more interesting, the possibility of using imaginative virtual reality computer games would be investigated. One option would be for Mrs PJ to wear a glove in which amplitude, speed and fractionation of movement are monitored with infrared sensors. Visual and auditory feedback are delivered online via a personal computer. One and a half hours of this type of training per day for two weeks has been found to have good effects (Jack et al. 2001 **A**; Merians et al. 2002 **A**). Another possibility would be to use robotic training devices for the hand (Hesse et al. 2005 **A**) or shoulder and elbow (Aisen et al. 1997 **A**; Volpe et al. 2000 **A**). One randomised controlled study found that practice of supination/pronation and wrist extension/flexion in a robotic device (comprising 800 repetitions over 20 minutes each working day, for six weeks) in addition to the usual in-patient physiotherapy, resulted in a better impairment and motor power outcome than electrical stimulation delivered over the same period of time. The total practice of upper limb tasks would be between 30 minutes and two hours per day, plus time spent in stretch positions.

The intensity and number of repetitions would be considered in the light of whether the aim of the immediate practice was to promote motor learning, improve endurance of a particular muscle/combination of muscles, or improve strength of a muscle/combination of muscles. Mrs PJ clearly had learning requirements, but would also have atrophic muscles (Ryan et al. 2002 **A**) and was likely to have had a gradual change to faster contractile motor unit properties (Gracies 2005 **R**) and therefore could have decreased strength and endurance. A functional repetition maximum (RM) would be determined to increase strength, using the exercise of lifting a heavy object by forward flexion in sitting (within pain free range) (Carr & Shepherd 2003b **C**). Between three and nine RM is sufficient to induce strength gains (Berger 1962 **B**), so to start with six RM would be used, that is, the weight Mrs PJ can lift six times and no more. Strength training may be carried out without adverse effects on muscle tone (Patten et al. 2004 **R**). Endurance can be increased by using low contraction force and sustaining and repeating the exercise (Richardson & Jull 1995 **C**), so active shoulder flexion movements in sitting while holding a lighter object, to a range between 30° and 70° forward flexion, holding for 10 seconds at end point, would be carried out.

Blocked practice would be used for a short time in the very early stages of training, to allow Mrs PJ to understand the requirements of the task (Landin & Herbert 1997 **B**). This would be followed by random practice, where exercises for forward flexion, abduction of the thumb, and using the fork would be practised in a random order, minimising consecutive repetitions of any one task (Schmidt & Wrisberg 2000 **C**). Random practice has been found to be superior to blocked practice for stroke patients learning a functional upper limb task involving reaching (Hanlon 1996 **A**). Table 4.4 shows how the order of exercises could potentially be constructed, though the actual order would be adjusted to performance. The schedule is subject to change according to the therapist's continued problem solving once practice has begun. The table does not list exercises for using the fork but these would be added into blocked and random practice sessions.

Practice at home would be organised to ensure the best chance of success. Firstly, the performance of the practice exercise(s) would be checked at the end of the one-to-one session with the therapist. A practice workbook would be issued to Mrs PJ, containing instructions for the exercise and tables to complete indicating the number of repetitions performed. A Polaroid might be pasted in the book to illustrate the desired movement. Key kinematic deviations to avoid, for example trunk flexion compensating for lack of forward flexion and elbow extension (Cirstea & Levin 2000 **A**), would be recorded with the exercise in the book. A check would be made to ensure Mrs PJ had the appropriate equipment/objects at home to do the exercise. Mrs PJ's performance would be checked with the therapist first thing next training session.

Variation of practice

Upper limb function involves many different goal-movement combinations, and even reaching in front itself may be performed under many varying conditions. It is imposs-ible to practise every single version of the reach sufficiently, so the learner must act as a problem solver, working out the appropriate movement for each new situation. Therefore, the exercises described above would be varied in one or more movement parameter to enable Mrs PJ to practise these problem solving skills. Such parameters could include movement speed, direction, or amplitude, and the object to be grasped, the immediate environment, or the final goal of the movement (for example, a cup

Table 4.4. Potential practice schedule for Weeks 1 and 2. Bullet points indicate the exercises practised

WEEK 1	Monday Outpatient appointment	Tuesday Home	Wednesday Home	Thursday Outpatient appointment	Friday Home	Saturday Home	Sunday Home
Motor Skill Learning	Blocked practice (30 minutes): • moving cup out • drawing arc • sliding down broom • getting ready • reach • holding ball • reach to can	Blocked practice (30 minutes): • same as Monday	Blocked practice (30 minutes): • same as Monday	Blocked practice with prescribed variations (30 minutes): • moving cup out • drawing arc • sliding down broom • getting ready • reach • holding ball • reach to can Upper limb group: • reach • other reach-to-grasp part and whole practice	Random practice with prescribed variations (30 minutes): • same as Thursday	Random practice with prescribed variations (30 minutes): • same as Thursday	Random practice with prescribed variations (30 minutes): • same as Thursday

(Continued)

Table 4.4. Potential practice schedule for Weeks 1 and 2. Bullet points indicate the exercises practised (*Continued*)

WEEK 1	Monday Outpatient appointment	Tuesday Home	Wednesday Home	Thursday Outpatient appointment	Friday Home	Saturday Home	Sunday Home
Stretch Positions	Supervised by therapist: • in external rotation • in protraction with thumb splint	Supervised by husband: • in external rotation • in protraction with thumb splint	Supervised by husband: • in external rotation • in protraction with thumb splint		Supervised by husband: • in external rotation • in protraction with thumb splint	Supervised by husband: • in external rotation • in protraction with thumb splint	
Strength and Endurance	6 RM forward flexion (3 sets)	10 × 10 second hold forward flexion and external rotation	6 RM forward flexion (3 sets)	10 × 10 second hold forward flexion and external rotation	6 RM forward flexion (3 sets)	10 × 10 second hold forward flexion and external rotation	

(*Continued*)

Table 4.4. Potential practice schedule for Weeks 1 and 2. Bullet points indicate the exercises practised (*Continued*)

	Monday Outpatient appointment	Tuesday Home	Wednesday Home	Thursday Outpatient appointment	Friday Home	Saturday Home	Sunday Home
WEEK 2							
Motor Skill Learning	Blocked practice (30 minutes): • starting off • move cup forward • moving tray Random practice with variations for previous tasks (if still required): • moving cup out • drawing arc • sliding down broom • getting ready • reach • holding ball • reach to can	Blocked practice (30 minutes): • same as Monday Random practice with variations for previous tasks: • same as Monday	Blocked practice (30 minutes): • same as Monday Random practice with variations for previous tasks: • same as Monday	Blocked practice with prescribed variations (30 minutes): • all tasks Upper limb group: • reach • other reach-to-grasp part and whole practice	Random practice with prescribed variations (30 minutes): • same as Thursday	Random practice with prescribed variations (30 minutes): • same as Thursday	Random practice with prescribed variations (30 minutes): • same as Thursday

(*Continued*)

Table 4.4. Potential practice schedule for Weeks 1 and 2. Bullet points indicate the exercises practised (*Continued*)

WEEK 2	Monday Outpatient appointment	Tuesday Home	Wednesday Home	Thursday Outpatient appointment	Friday Home	Saturday Home	Sunday Home
Stretch Positions	Supervised by therapist: • in external rotation • in protraction with thumb splint	Supervised by husband: • in external rotation • in protraction with thumb splint	Supervised by husband: • in external rotation • in protraction with thumb splint		Supervised by husband: • in external rotation • in protraction with thumb splint	Supervised by husband: • in external rotation • in protraction with thumb splint	
Strength and Endurance	6 RM forward flexion (3 sets)	10 × 10 second hold forward flexion and external rotation	6 RM forward flexion (3 sets)	10 × 10 second hold forward flexion and external rotation	6 RM forward flexion (3 sets)	10 × 10 second hold forward flexion and external rotation	

could be grasped in order to move it or to drink from it). One example of how some of the above exercises would be varied is given below. The variations are chosen to enhance the desired performance of the movement rather than making performance less than normal or causing greater abnormal kinematic deviations. Variation would be introduced when the patient demonstrated that they could perform the required movement.

Moving cup out

At the end of the movement, the cup is released, which gives practice of abduction of the thumb.

Drawing arc

The starting position is in standing, with shoulder in 90° forward flexion (position and amplitude variation).

Starting off

An increase in speed or reach is requested, which will cause a higher correlation between grasp and transport components at the start of the movement in healthy subjects (Van Vliet 1998). Patients with parietal lesions (like Mrs PJ), in contrast, lack these higher correlations with faster movements (Van Vliet & Sheridan, submitted A). The hand will usually open wider for faster movements, to compensate for increased spatial variability. Patients with parietal stroke have demonstrated an ability to do this also, but they open wider than healthy subjects (Van Vliet & Sheridan, submitted A). Increasing the speed will allow practice of both these aspects of reaching.

Sliding down broom handle

Slide hand up the broom handle, gripping the handle to pause the flexion when necessary (type of muscle contraction, eccentric, has changed to concentric, more difficult but more task-specific). Alternatively, perform in standing.

Getting ready to reach

Scapular setting can be incorporated into all the tasks described, at the start and end of each as required. Should difficulties be found in particular tasks, for example at the end of the forward flexion range, the principles of muscle imbalance can be adopted to identify which specific muscles are underactive, and exercises can be provided to specifically train that muscle in the range where the problem has been identified.

Moving cup forward

Arm is lifted slightly off table while cup is moved forward, which combines activation of anterior deltoid with serratus anterior and trapezius.

Reaching to cup

The cup is placed to require slight shoulder abduction (direction variation).

Reaching to can

The size of the object is varied (all large sizes) to allow practice of the ability to adjust the motor programme for different sizes of object. Patients with parietal stroke have demonstrated an ability to adjust grasp size in one study (Van Vliet & Sheridan, submitted **A**), although patients with a lesion of the intraparietal sulcus show a poor control of grasp aperture (Binkofsky et al. 1998 **A**).

FEEDBACK

Content of feedback

Studies have shown that patients with unilateral stroke are able to learn new motor skills (Hanlon 1996 **A**; Winstein et al. 1999 **A**), therefore Mrs PJ was expected to be able to learn as a result of practice. Her intrinsic feedback processes, which normally help to formulate the internal representation of the movement goal a person is trying to achieve, may have been compromised as a result of the stroke. Spatial perception and two-point discrimination were measurably impaired at assessment. Extrinsic feedback was therefore important for Mrs PJ. Boyd and Winstein (2001 **A**) have shown that implicit motor learning (learning perceptual motor skills by physical practice without conscious awareness) can be impaired in patients with stroke and so provision of knowledge of results (KR) may allow explicit memory (knowledge of facts, events and episodes) to assist motor learning (Winstein et al. 2005 **A**). Mrs PJ may also benefit from knowledge of performance (KP – 'information about the movement characteristics that led to the performance outcome' (Magill 2007 **C**)), since she does not have temporal lobe damage (such patients' implicit learning will be particularly affected (Boyd & Winstein 2001 **A**)).

Several of the prescribed practice tasks have inherent KR, for example, the patient will see when she has reached the target in 'moving cup out'. Other examples are 'drawing arc', 'moving cup forward' and 'moving tray'. When this occurs, additional KR may be redundant. Platz et al. (2001 **A**) examined the effect of KR in stroke subjects who were randomised into three groups and underwent a three-week training programme of upper limb tasks. The first group received the training with KR, the second without KR and the third did not have training. Although the training itself produced significant results compared to no training, when performance was measured at the end of the three weeks, there was no substantial extra effect for KR. The tasks chosen had inherent information about the movement outcome, for example hitting targets with a stylus and placing objects on top of other objects, so that additional KR, which took the form of bar diagrams on a computer screen, did not enhance learning further. This is a similar result to a study of healthy volunteers where extra verbal KR was redundant when outcome information was inherent in a task (Beukers et al. 1992 **B**). When Mrs PJ practised this type of task, KR would not be given about movement outcome. It would however be given about the quality of the movement performance, especially in early attempts. For example, her arm could potentially have abducted in the exercise 'moving cup out', so she would need feedback about this compensatory strategy. The feedback about the arm abduction would not necessarily focus on the body part, as an external focus of attention could be induced.

Feedback would be given verbally, and visual feedback via video would be used occasionally, with cues to direct Mrs PJ's attention to specific errors initially, when she was in the early stages of learning a task (Kernodle & Carlton 1992 **B**; Rothstein & Arnold 1976 **B**). This would work well with the exercises 'reaching to cup' and 'reaching to can', where cues could be used to direct attention to errors such as using excessive shoulder abduction or internal rotation. Feedback would be prescriptive (Schmidt & Wrisberg 2000 **C**), describing the errors and suggesting how to correct them, rather than descriptive (Kernodle & Carlton 1992 **B**), just describing the errors. For example: 'instead of moving your arm sideways, try to put more effort into moving your arm forwards'.

Attentional focus

Attentional focus can be directed either through the use of the environment or verbally. Where feedback about the outcome of the task (KR) can be obtained from the environment, this induces an external focus of attention whilst using intrinsic feedback mechanisms. For example 'moving cup out', 'drawing an arc', 'starting off', 'sliding down the broom', 'moving cup forward' all involve stickers which provide information about whether the task was achieved. Additional verbal feedback can be provided about the quality of the performance. By doing so, Mrs PJ would benefit from gaining additional information that could be used to adapt the motor programme and might be motivational (Schmidt & Wrisberg 2000 **C**). For example, for the task of drawing the arc with speed variation, the feedback could be, 'that was a little slow'. This could be followed with an instruction such as, 'for the next five movements I would like you to draw the arc more quickly', which reintroduces an external focus of attention. For the 'moving cup forward' task, to gain shoulder protraction, the feedback could be, 'in the last movement your shoulder did not come far enough forwards', and this again should be followed up by providing an external focus instruction.

Where possible, attentional is best focused towards the task objects.

In the tasks where KR is not explicit, words can be used to communicate the outcome of the movement (EF) or the quality of the performance (IF). This would be useful for the whole part practice tasks and perhaps the scapular setting task, where KR may be difficult to see independently. The choice of whether to use EF or IF feedback would depend upon how well the movement pattern was performed.

Scheduling of feedback

Verbal feedback would not be given while the task was actually being performed. Numerous studies in healthy subjects have shown that although concurrent feedback may enhance performance during practice, in retention or transfer tests performance is usually worse compared to conditions where feedback was provided after the movement was completed (Park et al. 2000 **B**; Schmidt & Wulf 1997 **B**; VanderLinden et al. 1993 **B**; Winstein et al. 1996 **B**). To the authors' knowledge, there are no studies comparing concurrent feedback with feedback after the movement in patients with stroke, so it is uncertain as to whether this finding in healthy subjects is also true for stroke subjects. The tasks practised in some of the studies with healthy subjects are similar to the tasks being practised here, however (Schmidt and Wulf, 1997 **B**; Winstein et al. 1996 **B**), so the findings are being cautiously applied. Similarly, because feedback that is delayed for several seconds after the movement is completed is demonstrably better than feedback given immediately after the movement in healthy subjects (Swinnen et al. 1990 **B**), the feedback given to Mrs PJ would be delayed for a few seconds. The explanation for these results is that both concurrent feedback and feedback immediately after the movement may prevent spontaneous error estimations, and encourage a dependency on extrinsic feedback (Van Vliet & Wulf 2006 **R**). Regarding concurrent feedback, an exception would be made for practice using virtual reality computer games, which typically include on-line feedback as part of the design.

Mrs PJ would not receive feedback on every attempt of a task, in order to encourage self-evaluation via the patient's own intrinsic feedback processes and greater movement stability (Salmoni et al. 1984 **B**; Schmidt 1991 **B**). Two studies of stroke patients and one of brain-injured patients demonstrate that a reduced feedback frequency can lead to better retention of a task. The two most relevant studies to this case, in which the subjects were learning an arm lever positioning task (Winstein et al. 1999 **A**) and a linear arm positioning task (Thomas & Harro 1996 **B**), found that feedback on 60 % of attempts led to better consistency of performance than feedback on 100 % (in the first study), and better movement accuracy with either 33 % or 67 % compared to 100 % (in the second).

Summary, average or bandwidth feedback would be used to reduce feedback frequency. In summary and average feedback the learner is given feedback about a set of trials (for example, five) after the set is completed. Where summary feedback involves feedback about every trial, average feedback refers to the average performance on that set of trials. Bandwidth feedback is given only when performance error exceeds a certain tolerance level (Schmidt & Wrisberg 2000 **C**). There are two papers showing support for these in brain-injured and stroke patients. One study by Croce, Horvat and Roswal (1996 **A**), using a coincidence timing task, provides some evidence for the effectiveness of summary and average feedback in individuals with traumatic brain injury. Compared to groups that received no feedback (control) or feedback after every trial, both summary and average feedback groups performed more effectively on an immediate retention test, and the summary feedback group was most accurate on a 24 hour retention test. A quasi-randomised study has examined the effect of kinematic feedback via electrogoniometry for the purpose of limiting knee hyperextension (Morris et al. 1992 **A**). Peak knee hyperextension was improved more than in the control group after a four week training period. The patients only received feedback if the knee was extended past the $0°$ position (bandwidth feedback), but there was no comparison to non-bandwidth feedback.

Several examples are now given of how summary, average and bandwidth feedback would be used with the prescribed exercises. Bandwidth feedback could be used for the 'starting off' exercise. The therapist would use visual observation to judge whether the hand opening had begun before the hand had moved forward a distance of 5 cm. If the therapist said nothing, the patient would know that their performance was adequate. The 5 cm distance could be reduced to encourage the temporal coupling to be tighter. Average feedback could be used for 'holding ball'. A tape measure would be used to ascertain the distance moved around the ball on each attempt. After 10 attempts, an average distance score would be communicated to the patient. Summary feedback would work well for 'sliding down broom handle'. The initial goal would be to increase the time taken to slide down to a certain point from the start position, in order to increase eccentric work of the shoulder flexors. A target time would be set, and at the end of a set of attempts, the number of attempts which took at least the target time would be communicated to the patient.

Wulf and Shea (2002 **B**; 2004 **B**; Wulf et al. 2002 **B**) caution that the learning of relatively complex skills might not benefit from, and might even be degraded by increasing the demands imposed on the learner by, for example, reducing the frequency of feedback. Some of the tasks above might be seen as complex for a stroke patient (although they are easy enough for a healthy person), so the response of Mrs PJ to the reduced feedback frequency would be closely monitored, and the frequency increased if necessary.

PRACTICE AND FEEDBACK IN LATER STAGES OF LEARNING

Once Mrs PJ could perform a skill as required, and showed some consistency of performance, random and varied practice would be introduced. An example of the timing of this introduction is shown in Table 4.4.

Feedback could become more precise (Gentile 1987 **C**), for example, she could receive feedback on the number of degrees of movement in 'moving cup out' or the number of millimetres moved in 'holding ball'. The frequency of feedback could be further reduced and when summary feedback was used, the number of attempts before feedback was given could be increased (Guadagnoli et al. 1996 **B**; Schmidt et al. 1990 **B**; Yao et al. 1994 **B**). If video was used in the later stages, self-evaluation would be encouraged as this works better for more experienced learners (Herbert et al. 1998 **B**).

ACKNOWLEDGEMENTS

The authors are very grateful to Mrs PJ, and to Frederike van Wijck and Mark Smith for their helpful comments on an earlier version of this chapter.

REFERENCES

Ada L, Goddard E, McCully J, Stavrinos T, Bampton J (2005) Thirty minutes of positioning reduces the development of shoulder external rotation contracture after stroke: a randomized
controlled trial. *Archives of Physical Medicine and Rehabilitation* **86**: 230–234.

Aisen ML, Krebs I, Hogan N, McDowell F, Volpe BT (1997) The effect of robot-assisted therapy
and rehabilitative training on motor recovery following stroke. *Archives of Neurology* **54**:
443–446.

Basmajian JV (1976) *Primary Anatomy* Williams and Wilkins.

Berger RA (1962) Optimum repetitions for the development of strength. *Research Quarterly
for Exercise and Sport* **33**: 334.

Beukers MA, Magill RA, Hall KG (1992) The effect of knowledge of results on skill acquisition
when augmented information is redundant. *Quarterly Journal of Experimental Psychology*
44A: 105–117.

Binkofsky F, Dohle C, Posse S, Stephan KM, Hefter H, Seitz RJ et al. (1998) Human anterior intraparietal area subserves prehension. A combined lesion and functional magnetic
resonance imaging activation study. *Neurology* **50**: 1253–1259.

Blair C (1995) Combining behaviour management and mutual goal setting to reduce physical
dependency in nursing home residents. *Nursing Research* **44**: 160–165.

Blair C, Lewis R, Vieweg V, Tucker R (1996) Group and single subject evaluation of a programme to promote self care in elderly nursing home residents. *Journal of Advanced
Nursing* **24**: 1207–13.

Blennerhassett J, Dite W (2004) Additional task-related practice improves mobility and upper
limb function early after stroke: a randomised controlled trial. *Australian Journal of
Physiotherapy* **50**: 219–224.

Bohannon RW, Smith MB (1987) Interrater reliability of a Modified Ashworth Scale of muscle
spasticity. *Physical Therapy* **2**: 206–207.

Boyd L, Winstein CJ (2001) Implicit motor-sequence learning in unilateral stroke: impact of
practice and explicit knowledge. *Neuroscience Letters* **298**: 65–69.

Braun C, Heinz U, Schweizer R, Wiech K, Birbaumer N, Topka H (2001) Dynamic organization
of the somatosensory cortex induced by motor activity. *Brain* **124**: 2259–2267.

Butefisch C, Hummelsheim H, Denzler P, Mauritz K-H (1995) Repetitive training of isolated
movements improves outcome of motor rehabilitation of the centrally paretic hand. *Journal of Neurological Sciences* **130**: 59–68.

Canning CG (2005) The effect of directing attention during walking under dual-task conditions
in Parkinson's disease. *Parkinsonism and Related Disorders* **11**: 95–99.

Carr J, Shepherd RB (2003a) Reaching and manipulation. In: *Stroke Rehabilitation: guidelines
for exercise and training to optimize motor skill* London: Butterworth Heinemann, pp.
159–206.

Carr JH, Shepherd RB (2003b) *Stroke Rehabilitation: guidelines for exercise and training to
optimize motor skill: scientific and evidence based exercise and training* London: Butterworth Heinemann.

Castiello U, Bennett KMB, Stelmach GE (1993a) The bilateral reach to grasp movement.
Behavioural Brain Research **56**: 43–57.

Castiello U, Bennett KMB, Stelmach GE (1993b) Reach to grasp: the natural response to
perturbation of object size. *Experimental Brain Research* **94**: 163–178.

Cirstea MC, Levin MF (2000) Compensatory strategies for reaching in stroke. *Brain* **123**:
940–953.

Cole JH, Furness AL, Twomey LT (1988) *Muscles in Action: an approach to manual muscle testing* Melbourne: Churchill Livingstone.

Creswell AG, Oddsson L, Thorstensson A (1994) The influence of sudden perturbations on trunk muscle activity and intra-abdominal pressure while standing. *Experimental Brain Research* **98**: 336–341.

Croce R, Horvat M, Roswal G (1996) Augmented feedback for enhanced skill acquisition in individuals with traumatic brain injury. *Perceptual and Motor Skills* **82**: 507–514.

Cunningham CL, Stoykov MEP, Walter CB (2002) Bilateral facilitation of motor control in chronic hemiplegia. *Acta Psychologia* **110**: 321–337.

Dijkermann HC, Letswaart M, Johnston M, MacWalter RS (2004) Does motor imagery training improve hand function in chronic stroke patients? A pilot study. *Clinical Rehabilitation* **18**: 538–549.

Falla D, Jull G, Hodges PW (2004) Feedforward activity of the cervical flexor muscles during voluntary arm movements is delayed in chronic neck pain. *Experimental Brain Research* **157**: 43–48.

Fasoli SE, Trombly CA, Ticle-Degned L, Verfaellie MH (2002) Effect of instructions on functional reach in persons with and without cerebrovascular accident. *American Journal of Occupational Therapy* **56**: 380–390.

Feys H, Weerdt WD, Verbeke G, Steck GC, Capiau C, Kiekens C et al. (2004) Early and repetitive stimulation of the arm can substantially improve the long-term outcome after stroke: a 5-year follow-up study of a randomized trial. *Stroke* **35**: 924–929.

Gentile AM (1987) Skill acquisition: action, movement and neuromotor processes. In: Carr JH, Shepherd RB (eds) *Movement Science: foundations for physical therapy in rehabilitation* London: Heinemann Physiotherapy.

Gentilucci M, Castiello U, Corradin ML, Scarpa M, Umilta C, Rizzolati G (1991) Influence of different types of grasping on transport component of prehension movements. *Neuropsychologica* **29**: 361–378.

Gracies J-M (2005) Pathophysiology of spastic paresis: 1: paresis and soft tissue changes. *Muscle and Nerve* **31**: 535–551.

Guadagnoli MA, Dornier LA, Tandy RD (1996) Optimal length for summary knowledge of results: the influence of task-related experience and complexity. *Research Quarterly for Exercise and Sport* **67**: 239–248.

Hanlon RE (1996) Motor learning following unilateral stroke. *Archives of Physical Medicine and Rehabilitation* **77**: 811–815.

Hawkins RJ, Mohtadi NGH (1991) Controversy in anterior shoulder instability. *Clinical Orthopaedics and Related Research* **272**: 152–161.

Hengeveld E, Banks K (2005) *Maitland's Peripheral Manipulation* London: Butterworth Heinemann.

Herbert E, Landin D, Menickelli J (1998) Videotape feedback: what learners see and how they use it. *Journal of Sport Pedagogy* **4**: 12–28.

Hesse S, Werner C, Pohl M, Rueckriem S, Mehrholz J, Lingau ML (2005) Computerized arm training improves the motor control of the severely affected arm after stroke. *Stroke* **36**: 1960–1966.

Higgs J, Jones M (2000) Clinical reasoning in the health professions. In: Higgs J, Jones M (eds) *Clinical reasoning* Oxford: Butterworth Heinemann, pp. 3–14.

Hodges P, Richardson C (1999) Altered trunk muscle recruitment in people with low back pain with upper limb movement at different speeds. *Archives of Physical Medicine and Rehabilitation* **80**: 1005–1012.

Hoff B, Arbib MA (1993) Models of trajectory formation and temporal interaction of reach and grasp. *Journal of Motor Behaviour* **25**: 175–192.

Hummelsheim H, Amberger S, Mauritz KH (1996) The influence of EMG-initiated electrical muscle stimulation on motor recovery of the centrally paretic hand. *European Journal of Neurology* **3**: 245–254.

Hummelsheim H, Hauptmann B, Neumann S (1995) Influence of physiotherapeutic facilitation techniques on motor evoked potentials in centrally paretic hand extensor muscles. *Electroencephalography and Clinical Neurophysiology* **97**: 18–28.

Hummelsheim H, Maier-Loth ML, Eickhof C (1997) The functional value of electrical muscle stimulation for the rehabilitation of the hand in stroke patients. *Scandinavian Journal of Rehabilitation Medicine* **29**: 3–10.

Jack D, Boian R, Merians AS, Tremaine M, Burdea GC, Adamovich SV et al. (2001) Virtual reality-enhanced stroke rehabilitation. *IEEE Transactions on Rehabilitation Engineering* **9**: 308–318.

Jackson J, Crow L (1991) The reliability of sensory assessments in hemiplegia. *Eleventh Congress of the World Confederation for Physical Therapy* 514–516.

Jeannerod M (1994) The representing brain: neural correlates of motor intention and imagery. *Behavioural and Brain Sciences* **17**: 187–245.

Joynt RL (1992) The source of shoulder pain in hemiplegia. *Archives of Physical Medicine and Rehabilitation* **58**: 409–413.

Kennerley SW, Sakai K, Rushworth MFS (2003) Organization of action sequences and the role of the pre-SMA. *Journal of Neurophysiology* **91**: 978–993.

Kernodle MW, Carlton LG (1992) Information feedback and the learning of multiple-degree-of-freedom activities. *Journal of Motor Behaviour* **24**: 187–196.

Kumar R, Metter EJ et al. (1990) Shoulder pain in hemiplegia: the role of exercise. *American Journal of Physical Medicine and Rehabilitation* **69**: 205–208.

Landers M, Wulf G, Wallman H, Guadagnoli MA (2005) An external focus of attention attenuates balance impairment in Parkinson's disease. *Physiotherapy* (In press).

Landin D, Herbert EP (1997) A comparison of three practice schedules along the contextual interference continuum. *Research Quarterly for Exercise and Sport* **68**: 357–361.

Lincoln NB, Leadbitter D (1979) Assessment of motor function in stroke patients. *Physiotherapy* **65**: 48–51.

Lotze M, Braun C, Birbaumer N, Anders S, Cohen LG (2003) Motor learning elicited by voluntary drive. *Brain* **126**: 866–872.

Luft AR, McCombe-Waller A, Whitall J, Forrester LW, Macko R, Sorkin JD et al. (2004) Repetitive bilateral arm training and motor cortex activation in chronic stroke. *Journal of the American Medical Association* **292**: 1853–1861.

Magill RA (2007) *Motor Learning and Control: concepts and applications* (8 edition) New York: McGraw-Hill.

McNevin N, Wulf G, Carlson C (2000) Effects of attentional focus, self-control, and dyad training on motor learning: implications for physical rehabilitation. *Physical Therapy* **80**: 373–385.

McNevin NH, Shea CH, Wulf G (2003a) Postural control changes in cva patients a function of supra-postural attentional focus. *Psychological Research* **67**: 22–29.

McNevin NH, Wulf G (2002) Attentional focus on supra-postural tasks affects postural control. *Human Movement Science* **21**: 187–202.

McNevin NH, Wulf G (2003b) Increasing the distance of an external focus of attention enhances learning. *Medicine and Science in Sports and Exercise* **35**: S315.

Melzack R (1987) The short-form McGill pain questionnaire. *Pain* **30**: 191–197.

Merians AS, Jack D, Boian R, Tremaine M, Burdea GC, Adamovich SV et al. (2002) Virtual reality-augmented rehabilitation for patients following stroke. *Physical Therapy* **82**: 898–915.

Mimori K, Muneta T, Nakagawa T, Shinomiya K (1999) A new pain provocation test for superior labral tears of the shoulder. *American Journal of Sports Medicine* **27**: 137–142.

Morris ME, Matyas TA, Bach TM, Goldie PA (1992) Electrogoniometric feedback: its effect on genu recurvatum in stroke. *Archives of Physical Medicine and Rehabilitation* **73**: 1147–1154.

Mottram SL (1997) Dynamic stability of the scapula. *Manual Therapy* **2**: 123–131.

Mudie MH, Matyas TS (1996) Upper extremity retraining following stroke: effects of bilateral practice. *Journal of Neurological Rehabilitation* **10**: 167–184.

Napier JR (1956) The prehensile movements of the human hand. *Journal of Bone and Joint Surgery* **38B**: 902–913.

Neer CS (1972) Anterior acromioplasty for the chronic impingement syndrome in the shoulder. A preliminary report. *Journal of Bone and Joint Surgery* **54A**: 41–50.

Nordin M, Frankel VH (2001) *Basic Biomechanics of the Musculoskeletal System* Philadelphia: Lippincott Williams and Wilkins.

Nouri FM, Lincoln NB (1987) An extended activities of daily living scale for stroke patients. *Clinical Rehabilitation* **1**: 301–305.

Page SJ, Levine P, Leonard AC (2005) Effects of mental practice on affected limb use and function in chronic stroke. *Archives of Physical Medicine and Rehabilitation* **86**: 399–402.

Park J-H, Shea CH, Wright DL (2000) Reduced frequency concurrent and terminal feedback: a test of the guidance hypothesis. *Journal of Motor Behaviour* **32**: 287–296.

Partridge C, Johnston M (1989) Perceived control of recovery from physical disability: measurement and prediction. *British Journal of Clinical Psychology* **28**: 53–59.

Patten C, Lexell J, Brown HE (2004) Weakness and strength training in persons with post-stroke hemiplegia: rationale, method, and efficacy. *Journal of Rehabilitation Research and Development* **41**: 293–312.

Platz T, Winter T, Muller N, Pinkowski C, Eickhof C, Mauritz K-H (2001) Arm ability training for stroke and traumatic brain injury patients with mild arm paresis: a single-blind, randomized, controlled trial. *Archives of Physical Medicine and Rehabilitation* **82**: 961–968.

Rasch PJ, Morehouse CE (1957) Effect of static and dynamic exercises on muscular strength and hypertrophy. *Journal of Applied Physiology* **11**: 129–134.

Reid A, Chesson R (1998) Goal attainment scaling. Is it appropriate for stroke patients and their physiotherapists? *Physiotherapy* **84**: 136–144.

Rey A (1959) Le test, de copie de figure complexe. Paris: Editions Centre de Psychologie Applique.

Richardson CA, Jull GA (1995) Muscle control – pain control. What exercises would you prescribe? *Manual Therapy* **1**: 1–9.

Rothstein AL, Arnold RK (1976) Bridging the gap: application of research on videotape feedback and bowling. *Motor Skills: Theory into Practice* **1**: 36–61.

Rutherford OM (1988) Muscular coordination and strength training: implications for injury rehabilitation. *Sports Medicine* **5**: 196–202.

Ryan AS, Dobrovny CL, Smith GV, Silver KH, Macko RF (2002) Hemiparetic muscle atrophy and increased intramuscular fat in stroke patients. *Archives of Physical Medicine and Rehabilitation* **83**: 1703–1707.

Sale JB, MacDougall D (1981) Specificity of strength training: a review for the coach and athlete *Canadian Journal of Applied Sports Sciences* **6**: 87–92.

Salmoni AW, Schmidt RA, Walter CB (1984) Knowledge of results and motor learning: a review and critical reappraisal. *Psychological Bulletin* **95,** 355–386.

Schmidt RA (1991) Frequent augmented feedback can degrade learning: evidence and interpretations. In: Requin J, Stelmach GE (eds) *Tutorials in Motor Neuroscience* Dordrecht: Kluwer Academic Publishers, pp. 59–75.

Schmidt RA, Lange C, Young DE (1990) Optimizing summary knowledge of results for skill learning. *Human Movement Science* **9**: 325–348.

Schmidt RA, Wrisberg CA (2000) *Motor Learning and Performance: a problem-based learning approach* Champaign, Illinois: Human Kinetics.

Schmidt RA, Wulf G (1997) Continuous concurrent feedback degrades skill learning: implications for training and simulation. *Human Factors* **39**: 509–525.

Shea CH, Wulf G (1999) Enhancing motor learning through external-focus instructions and feedback. *Human Movement Science* **18**: 553–571.

Sunderland A, Tinson D, Bradley L, Fletcher D, Hewer RL, Wade DT (1992) Enhanced physical therapy for arm function after stroke. *Journal of Neurology, Neurosurgery and Psychiatry* **55**: 530–535.

Swinnen S, Schmidt RA, Nicholson DE, Shapiro DC (1990) Information feedback for skill acquisition: instantaneous knowledge of results degrades learning. *Journal of Experimental Psychology: learning, memory and cognition* **16**: 706–716.

Thomas DM, Harro CC (1996) Effects of relative frequency of knowledge of results on brain injured and control subjects learning a linear positioning task. *Neurology Report* **20**: 60–62.

Tuller B, Turvey MT, Fitch HL (1982) The Bernstein perspective II. The concept of muscle linkage or co-ordinative structure. In: Kelso JAS (ed.) *Human Motor Behaviour* Hillsdale, New Jersey: Lawrence Erlbaum Publishers.

Tyc F, Boyadjian A, Devanne H (2005) Motor cortex plasticity induced by extensive training revealed by transcranial magnetic stimulation in human. *European Journal of Neuroscience* **21**: 259–266.

VanderLinden DW, Cauraugh JH, Greene TA (1993) The effect of frequency of kinetic feedback on learning an isometric force production task in nondisabled subjects. *Physical Therapy* **73**: 79–87.

Van Vliet P (1988) Kinematic analysis of videotape to measure walking following stroke: a case study. *Australian Journal of Physiotherapy* **34**: 48–51.

Van Vliet P, Kerwin DG, Sheridan MR, Fentem PH (1995) The influence of functional goals on the kinematics of reaching following stroke. *Neurology Report* **19**: 11–16.

Van Vliet PM (1998) *An Investigation of Reaching Movements Following Stroke* PhD thesis, Nottingham.

Van Vliet PM, Sheridan MR Coordination between reaching and grasping in patients with hemiparesis and normal subjects. (Submitted for publication).

Van Vliet PM, Wulf G (2006) Extrinsic feedback for motor learning after stroke: what is the evidence? *Disability and Rehabilitation* **28**(13–14): 831–840.

Vattanasilp W, Ada L, Crosbie J (2000) Contribution of thixotrophy, spasticity, and contracture to ankle stiffness after stroke. *Journal of Neurology, Neurosurgery and Psychiatry* **69**: 34–39.

Volpe BT, Krebs HI, Hogan N, Edelstein OTR, Diels C, Aisen M (2000) A novel approach to stroke rehabilitation: robot-aided sensorimotor stimulation. *Neurology* **54**: 1983–1944.

Waller SM, Harris-Love M, Liu W, Whitall J (2006) Temporal coordination of the arms during bilateral simultaneous and sequential movements in patients with chronic hemiparesis. *Experimental Brain Research* **168**: 450–454.

Whitall J, Waller SM, Silver KHC, Macko RF (2000) Repetitive bilateral arm training with rhythmic auditory cueing improves motor function in chronic hemiparetic stroke. *Stroke* **31**: 2390–2395.

Wilson BA, Cockburn J, Baddeley A (1985) *The Rivermead Behavioural Memory Test* Bury St Edmonds: Thames Valley Test Company.

Winstein C, Wing AM, Whitall J (2005) Motor control and learning principles for rehabilitation of upper limb movements after brain injury. In: Grafman J (ed.) *Handbook of Neuropsychology*, Vol. 9 Amsterdam: Elsevier.

Winstein CJ, Merians AS, Sullivan KJ (1999) Motor learning after unilateral brain damage. *Neuropsychologia* **37**: 975–987.

Winstein CJ, Pohl PS, Cardinale C, Green A, Scholtz L, Waters CS (1996) Learning a partial-weight-bearing skill: effectiveness of two forms of feedback. *Physical Therapy* **76**: 985–993.

Winstein CJ, Rose DK, Tan SM, Lewthwaite R, Chui HC, Azen SP (2004) A randomized controlled comparison of upper-extremity rehabilitation strategies in acute stroke: a pilot study of immediate and long-term outcomes. *Archives of Physical Medicine and Rehabilitation* **85**: 620–628.

Wulf G, Hob M, Prinz W (1998) Instructions for motor learning: differential effects of internal versus external focus of attention. *Journal of Motor Behaviour* **30**: 169–179.

Wulf G, Lauterbach B, Toole T (1999) Learning advantages of an external focus of attention in golf. *Research Quarterly for Exercise and Sport* **70**: 120–126.

Wulf G, McConnel N, Gartener M, Schwarz A (2002) Enhancing the learning of sports skills through external-focus feedback. *Journal of Motor Behaviour* **34**: 171–182.

Wulf G, Shea C, Park JH (2001a) Attention in motor learning: preferences for and advantages of an external focus. *Research Quarterly for Exercise and Sport* **72**: 335–344.

Wulf G, Shea CH (2002) Principles derived from the study of simple motor skills do not generalize to complex skill learning. *Psychometric Bulletin and Review* **9**: 185–211.

Wulf G, Shea CH (2004) Understanding the role of augmented feedback: the good, the bad, and the ugly. In: Williams AM, Hodges NJ (eds) *Skill Acquisition in Sport: research, theory and practice* London: Routledge, pp. 121–144.

Wulf G, Shea CH, Park JH (2001b) Attention and motor performance: preferences for and advantages of an external focus. *Research Quarterly for Exercise and Sport* **72**: 335–344.

Wulf G, Weigelt C (1997) Instructions about physical principles in learning a complex motor skill: to tell or not to tell. *Research Quarterly for Exercise and Sport* **68**: 362–367.

Yao W-X, Fischman MG, Wang YT (1994) Motor skill acquisition and retention as a function of average feedback, summary feedback, and performance variability. *Journal of Motor Behaviour* **26**: 273–282.

Zachry T, Wulf G, Mercer J, Bezodis N (2005) Increased movement accuracy and reduced EMG activity as a result of adopting an external focus of attention. *Brain Research Bulletin* **67**: 304–309.

5 Improving Walking After Stroke Using a Treadmill

LOUISE ADA AND CATHERINE M. DEAN

CASE REPORT I

BACKGROUND

Mrs PG is 65 and lives with her husband, who is still working part time. She woke up not being able to speak coherently and not being able to move the right side of her body. The ambulance was called and she was admitted to hospital. It is now Day Six.

MEDICAL STATUS

Diagnosed having had a stroke. Conscious. On blood pressure lowering medication.

IMPAIRMENTS

Weakness – severe in most lower limb muscles and all upper limb muscles.
Incoordination – unable to be assessed due to severe weakness.
Spasticity – no spasticity – Tardieu scale score X = 0 at V3 (fast velocity) during ankle dorsiflexion and elbow extension.
Sensation – normal.
Language – expressive aphasia so she understands 90 % but can only communicate about 40 % of what she wants to say.
Cognition – normal.
Perception – normal.

ACTIVITY LIMITATIONS

Standing – cannot stand independently, needs help from one person.
Walking – cannot walk independently, needs substantial help from two people.
Use of upper limb – no voluntary movement at any joint.

Recent Advances in Physiotherapy. Edited by C. Partridge
© 2007 John Wiley & Sons, Ltd

QUESTION 1

Should treadmill training with body weight support (BWS) be used to retrain walking?

The first step in answering this question is to decide on the outcomes of interest. Given that Mrs PG is non-ambulatory, whether or not treadmill training with BWS is effective at establishing walking will be of prime interest. Furthermore, the quality of the walking produced by the training will be of interest. There are numerous outcome measures which evaluate walking, ranging from performance-based tests (such as the 10-m Walk Test (Wade 1992)) to ordinal scales (such as Item 5 of the Motor Assessment Scale for Stroke (Carr et al. 1985)). The most commonly used measure in clinical trials is the 10-m Walk Test, probably because it is simple to carry out, reliable, and yields continuous data. Furthermore, the most common parameter reported is walking speed and, while not measuring quality of walking directly, it nevertheless reflects qualitative gait parameters such as step length and cadence. This relationship is described in the equation:

$$\text{Speed} = \frac{\text{stride length} \times \text{cadence}}{120}$$

Therefore, proportion of patients walking and the 10-m Walk Test are probably the best measures reflecting the outcomes of interest.

The next step in answering the question is to look for evidence of whether treadmill training improves the proportion of people walking independently, or the quality of walking. Considering the highest levels of evidence first, there are two systematic reviews assessing the efficacy of treadmill training with BWS after stroke. The efficacy of treadmill training with BWS was considered in a review by Van Peppen and colleagues (2004 **A**). They concluded that treadmill training with BWS does not improve walking speed or ability although it does appear to improve walking endurance. However, most of the participants in the trials included in this review were already walking and so this finding is of limited use in answering the question. The efficacy of treadmill training with BWS was also considered in a Cochrane review by Moseley and colleagues (2005 **A**). They did separate their analyses into those who were non-ambulatory versus those who were ambulatory. They report that there is no greater risk of being non-ambulatory or a dependent walker if treadmill training with BWS is used than if other more conventional interventions are used (RR 1.1, 95 % CI 0.9 to 1.3). This finding was based on 178 participants in five randomised trials (da Cunha Filho et al. 2002, Kosak et al. 2000, Nilsson et al. 2001, Scheidtmann et al. 1999, Werner et al. 2002 **A**). Furthermore, walking speed was no different as a result of the interventions (WMD –0.01 m/s, 95 % CI –0.08 to 0.06). This finding was based on 148 participants in four randomised trials (da Cunha Filho et al. 2002; Kosak et al. 2000; Nilsson et al. 2001; Werner et al. 2002 **A**). The more conventional therapy used in these trials was always exactly matched for frequency and duration and was usually carried out for 20–45 minutes, five days a week. Two trials used a motor learning approach (da Cunha Filho et al. 2002; Nilsson et al. 2001 **A**), while one trial used a neurophysiologic approach (Scheidtmann et al. 1999), one used an orthopaedic approach (Kosak et al. 2000 **A**), and one used another walking device (Werner et al. 2002 **A**).

Given that treadmill walking with BWS is no more or less effective than the same amount of conventional therapy, the decision of whether to undertake it with Mrs PG will have to be made on other factors. Such factors are the efficient use of staff time, the amount of practice likely to be undertaken during overground walking versus treadmill walking, and Occupational Health and Safety. At the moment, it takes two therapists to help Mrs PG practise the whole task of walking overground. While this was controlled in the randomised trials, it is unlikely two staff members will be free to help her for very long in ordinary clinical practice. Walking on the treadmill with BWS means that she may only need the help of one therapist to move her affected leg forward in swing phase, even if it takes two people to assist her onto the treadmill. It is likely to be easier to move Mrs PG's leg during swing when she is in one place on a treadmill than to support it during swing and stance while she is trying to progress overground (since it does not matter if the knee flexes during stance, as the body is supported). Treadmill walking with partial weight support via an overhead harness provides the opportunity to complete larger amounts of walking practice, for example, even if patients only walk for five minutes at a slow speed of 0.2 m/s supported on a treadmill, they will 'walk' 60 m (Crompton et al. 1999 **C**). It is likely, therefore, that Mrs PG will undertake more practice of the whole task of walking if she does treadmill training with BWS.

Taking into account all the evidence, treadmill training with BWS should be an intervention capable of establishing walking in Mrs PG.

QUESTION 2

How should treadmill training with BWS be applied to improve the likelihood of the patient becoming ambulatory with good quality of walking?

To answer this question, observational studies of treadmill and overground walking after stroke can be examined. These studies compare walking overground with walking on a treadmill with BWS in stroke patients who are just walking or walking with difficulty (Chen et al. 2005a, 2005b; Hassid et al. 1997; Hesse et al. 1997 **A**). One of the common findings is that by adding BWS, the symmetry of walking is improved, due to the increased time the affected leg spends in single stance phase. However, there may be a limit to how much support should be given. Hesse and colleagues (1997 **A**) compared 0, 15, 30, 45 and 60 % BWS. They found that over 30 % BWS resulted in markedly abnormal muscle activity in six lower limb muscles they examined. This has resulted in a maximum of 30 % BWS becoming something akin to an industry standard. Perhaps the most useful information comes from Chen and colleagues (2005a, 2005b **A**), who systematically varied BWS, speed of treadmill, stiffness of the support harness, and support from a handrail. They found that different factors were helpful in different aspects of walking. For example, increasing BWS combined with support from a handrail produced the most symmetrical walking in terms of time spent in single stance phase, whereas increasing speed increased energy at toe-off. Increasing the stiffness of the support harness increased energy cost during swing phase, which may be both good and bad. Sullivan and colleagues (2002 **A**) carried out a randomised trial comparing three treadmill speeds during training with BWS for patients who could walk but walked slowly. They found that the fastest treadmill speed increased final overground walking speed by 0.13 m/s (p = 0.02) more than the two slower speeds.

It makes sense to examine how the treadmill training with BWS was carried out for the non-ambulatory participants in the randomised trials used in the Cochrane review (da Cunha Filho et al. 2002; Kosak et al. 2000; Nilsson et al. 2001; Werner et al. 2002 **A**). The training was carried out for 20–45 minutes every weekday. All studies report manipulating BWS and treadmill speed to progress training. Initial BWS varied from 10–100 % across trials. Number of therapists assisting, whether support from a handrail was allowed, whether shoes were worn, and whether the ankle was splinted, were reported variably across trials. Perhaps the most specific information on the interaction between treadmill speed and BWS comes from da Cunha Filho and colleagues (2002 **A**). They report that BWS was started at 30 % and decreased until knee flexion during stance was no more than 15°. When normal step length could be taken consistently, the speed of the treadmill was increased incrementally, by 0.01 m/s at a time.

We have gained some additional insights into training non-ambulatory people after stroke through carrying out a large, multicentre randomised trial which is expected to be finished in mid 2007 (http://www.clinicaltrials.gov Identifier NCT00167531 **C**). Our experience during this trial suggests that attention should be directed to several areas – support of the patient, method of therapist assistance, and progression of training. If the patient is severely disabled, it is more efficient to apply the harness in lying, transfer them to the treadmill by wheelchair, and use the automatic lift function to lift them into standing, than to put the harness on in sitting and get them to stand up by themselves. If the affected arm has no voluntary muscle activity, use a firm sling to support it, but if there is some activity, put the hand to the handrail using a bandage or a weightlifting splint (see Figure 5.1). We have found metronomes to be useful in enhancing rhythmical stepping and thereby directing step length; for example, slowing the metronome down will result in alternate feet staying on the 'ground' for longer. The most difficult job for the therapist is to lift the affected leg through during swing phase (Figure 5.2a). When the leg is very weak, a length of theraband can be tied from the front of the shoe to the front bar of the treadmill, which will serve to pull the leg forward when the weight is released (Figure 5.2c). Alternately, the affected foot can be placed in a pillow slip and twisted at the front (Figure 5.2b) so that the foot can be lifted from the toe, thereby enhancing dorsiflexion of the ankle. The therapist can sit on a chair turned backwards, which will support the trunk, making lifting the affected leg easier. It is important that the therapist assists the leg only in swing phase, and encourages the patient to extend their lower limb during stance, allowing the BWS to prevent the patient collapsing. To progress the training, when step length is consistently normal, we increase the speed until step length is compromised. When

Figure 5.1. Using a splint to support the affected hand on the hand rail.

the knee can be held straight during stance phase, we reduce the BWS. We have found that an easy transition is made to overground walking when the patient can walk on the treadmill at 0.5 m/s with $\leq$ 10 % BWS.

PLAN: TO ESTABLISH GOOD QUALITY WALKING IN MRS PG

A specific intervention plan, based on the above evidence, to carry out treadmill training with BWS for Mrs PG using a treadmill and overgound BWS system, is outlined below:

Gain medical clearance and consent to participate in exercise programme

Consult with Mrs PG's treating doctor to organise medical clearance to participate in treadmill walking training with BWS. Put harness on in lying and make sure Mrs PG is wearing shoes. Apply triangular sling to affected arm. Wheel Mrs PG onto treadmill in a wheelchair. Use the automatic lift function to lift her into standing. Given that Mrs PG has communication problems, modified safety procedures will have to be put in place. Attach safety strap, have relative or aide standing by emergency stop switch and teach Mrs PG a signal to indicate that the treadmill should be stopped.

Initial treadmill and BWS programme

To begin with, do not run the treadmill. Allow Mrs PG to hang on to a handrail. Increase BWS to 30 % in standing and make sure knee of affected leg is bent no more than 15 degrees. If it is, increase BWS. Put Mrs PG's affected foot in a pillow slip

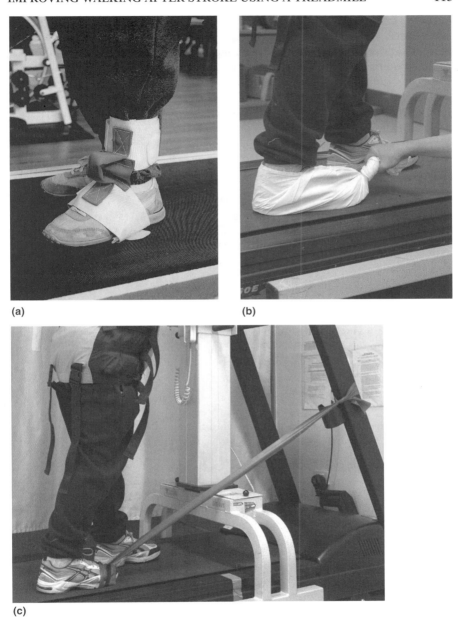

(a)

(b)

(c)

Figure 5.2. Using a) custom-made splint, b) pillowcase, and c) theraband to assist with lifting the affected leg forward during swing phase.

and twist at the front. Turn on a metronome at a frequency which matches the highest cadence that Mrs PG can manage. Sit on a low stool and help Mrs PG to walk on the spot in time with the rhythm by using the right hand to flex the knee and the left hand to lift the twisted part of the pillowcase. Then turn the treadmill on as slowly as possible. Mrs PG should keep walking in time with the metronome – the metronome frequency and the treadmill speed will determine her step length. Lift the affected leg forward during swing phase but encourage Mrs PG to extend her lower limb during stance and allow the BWS to hold her up. Count steps for encouragement and take a rest every two minutes at first.

Progressing treadmill and BWS programme

Increase step length by slowing down the metronome. When step length is increased, increase the speed until step length is compromised. When Mrs PG can straighten her knee from the 15 degrees, reduce the BWS. Continue to alternate these two strategies until she is walking at 0.5 m/s with $\leq 10\,\%$ BWS. At this stage begin to do overground walking with BWS.

Overground and BWS programme

Lock the wheels of the support frame so that it will only run in one direction. Put markers on the floor to increase step length and constrain step width. Apply only the trunk/pelvis part of harness, firmly. Push the support frame as Mrs PG walks forwards and then backwards overground. Progress by loosening the vertical support straps, getting Mrs PG to push the frame herself, and increasing step length and decreasing step width (see Figure 5.3).

Monitoring progression and enhancing compliance

At the beginning, record the number of steps to provide encouragement. Then, as ability improves, record distance covered on treadmill, highest speed and lowest amount of BWS – graph to provide motivation to improve. Record distance, step length and width during overground walking with BWS. As independent walking overground is possible, use 10-m Walk Test at the same time every week to monitor progress. As well as timing over the 10-m, count the number of steps and calculate average step length and cadence.

CASE REPORT II

BACKGROUND

Mr IB is 70 and lives alone. He has a very supportive daughter nearby, although she is busy bringing up four children. He suffered a stroke two years ago. Recently he has felt that his walking has deteriorated, and has approached a physiotherapy ambulatory care service for help.

Figure 5.3. Using a portable system to practise overground walking. Harness is for safety only. Markers on the floor encourage a long step length and narrow step width (refers to Case Report II).

MEDICAL STATUS

On blood pressure lowering medication.

IMPAIRMENTS

Weakness – moderately strong in lower limb muscles.

Incoordination – slight problem with incoordination in both upper limb and lower limb.

Spasticity – mild spasticity – Tardieu scale score $X = 1$ at V3 (fast velocity) during ankle dorsiflexion and $X = 2$ during elbow extension.

Contracture – loss of 10° ankle dorsiflexion.

Sensation – tactile and kinaesthetic sensation moderately impaired.

Language – normal.

Cognition – slight memory loss.

Perception – normal.

ACTIVITY LIMITATIONS

Standing – can stand with feet together and look over both shoulders without falling or having to take a step, but uses arms.

Walking – can walk independently, but very slowly and carefully at 0.6 m/s and 190 m in six min.

Use of upper limb – can use for support but not manipulation of objects (mostly due to loss of sensation).

QUESTION 1

Should treadmill training be used to improve community ambulation?

The first step in answering this question is to ascertain which of the commonly used walking outcome measures is the best indicator of community ambulation. There are numerous outcome measures which evaluate walking, ranging from performance based tests such as the 10-m Walk Test or 6-min. Walk Test, to ordinal scales such as Item 5 of the Motor Assessment Scale for Stroke, to self-reported questionnaires such as the Walking Impairment Questionnaire (Regensteiner et al. 1990). One commonly used performance based test is the 6-min. Walk Test, in which the distance covered in six minutes is recorded. The 6-min. Walk Test measures sustained effort and therefore reflects walking capacity, which is an essential component of community ambulation. Moreover, the 6-min. Walk Test has well documented standardised procedures and instructions, and there is normative data for persons aged between 40 and 80 years (Enright & Sherill 1998). Previous research has highlighted the shortcomings of using the 10-m Walk Test to predict walking capacity. Dean and colleagues (2001) measured 10-m Walk and 6-min. Walk Tests on healthy subjects and individuals after stroke, and found that using performance on the 10-m Walk Test to predict that on 6-min. Walk Test resulted in an overestimation of walking capacity. Therefore, of the commonly used walking outcome measures, the 6-min. Walk Test is likely to be the best predictor of community ambulation.

The next step is to look for evidence of whether treadmill training improves performance on the 6-min. Walk Test. Considering the highest levels of evidence first, there are two systematic reviews assessing the efficacy of treadmill training after stroke. Moseley and colleagues (2005 **A**) have completed a Cochrane review and reported the results of the review were not conclusive. There were no statistically significant differences between treadmill training, with or without body weight support, and other interventions on walking speed or dependence. Secondary analysis indicated that among people with stroke who could walk independently at the start of treatment, treadmill training may improve walking speed. Moseley and colleagues reported that the methodological quality of studies was poor and few studies reported 6-min. Walk Test. The efficacy of treadmill training with and without body weight support was also considered in a review by Van Peppen and colleagues (2004 **A**). They concluded that treadmill training with body weight support improved walking endurance and treadmill training without body weight support improved walking ability as measured on the Functional Ambulation Category (Wade 1992). One of the difficulties in analysing the data from these reviews is the difference in study design and methodological quality. Studies have included ambulatory and non-ambulatory individuals, acute and chronic individuals, and provided treadmill training with or without body weight support as well as other interventions. For example, the studies included in the Van Peppen review included individuals very early after stroke (10 days) as well as individuals 26 months after stroke.

Given that the evidence from the systematic reviews was in general supportive of treadmill training, the next step in answering our question is to examine the trials whose participants most closely reflect the characteristics of Mr IB, that is, someone who walks independently at about half the speed of his age-matched counterparts, two years after a stroke. Two randomised trials which examined individuals who were ambulatory after chronic stroke fit this criterion. Ada and colleagues (2003 **A**) examined the effect of a four week treadmill and overground walking programme, consisting of three 30 minute sessions a week, compared to a placebo of low intensity home exercises. Macko and colleagues (2005 **A**) examined the effect of six months of three 40 minute progressive aerobic (60–70 % heart rate reserve) treadmill sessions per week, compared to six months of three 35 minute sessions of supervised stretching, and five minutes of low intensity (30–40 % heart rate reserve) treadmill walking, per week. Both studies found a significant effect on walking capacity measured using the 6-min. Walk Test. The between-group effect size reported by Ada and colleagues immediately following the four week programme was 86 m (95 % CI 44 to 128), and three months later was 30 m (95 % CI 0 to 60). Macko and colleagues reported a between group effect of 43 m ($p = 0.02$). Ada and colleagues also reported a greater increase in walking speed and step length with treadmill and overground walking training compared with the sham intervention.

In addition to the trials that match Mr IB's characteristics, there is more evidence (although at a weaker level) which suggests treadmill walking may be a useful intervention to improve both the speed and capacity of walking in such patients. In uncontrolled trials of chronic stroke patients, treadmill walking has been associated with increases in strength (Smith et al. 1998, 1999 **A**), decreases in energy expenditure (Macko et al. 1997, 2001 **A**), as well as increases in walking speed and quality (Silver et al. 2000 **A**).

Taking into account all the evidence, treadmill training should be an intervention capable of improving Mr IB's community ambulation.

QUESTION 2

How should treadmill training be applied to improve community ambulation?

The most logical approach to answering this question is to examine how the intervention was implemented in the two trials which provided evidence that treadmill walking was effective in improving six minute distance (Ada et al. 2003; Macko et al. 2005 **A**). Macko and colleagues' training programme was six months, 10–40 minute sessions, three times a week. The sessions were characterised by progressive increases in duration (five minutes per session every two weeks) and in aerobic intensity (5 % Heart Rate Reserve every two weeks, achieved by increasing the speed of the treadmill). Training speed increased from 0.48 ± 0.3 m/s at baseline to 0.75 ± 0.3 m/s at six months, and training duration increased from 12 ± 6 minutes to 41 ± 10 minutes at six months.

Ada and colleagues implemented a training programme three times a week for only four weeks. The training sessions comprised 30 minutes of walking, which took about 45 minutes to accomplish. Each session consisted of both treadmill and overground walking, with the proportion of treadmill walking decreasing by 10 % each week, from 80 % in Week One to 50 % in Week Four. Subjects received individual training from a physical therapist; however, there was some opportunity for social interaction since two subjects were trained concurrently. The programme was carried out in a community setting and transport was provided if necessary. The treadmill walking component was structured to increase step length, speed, balance, fitness, and automaticity. To increase step length, the treadmill was run at a comfortable speed and instructions such as 'walk as slowly as possible' or 'take as few steps as possible' were used. When a normal step length was observed, the speed of the treadmill was increased (until step length was compromised). When maximum speed was achieved, balance was challenged by reducing the degree of hand support, and fitness encouraged by increasing the incline of the treadmill, thereby increasing workload. Finally, automaticity was promoted by presenting the subjects with a concurrent cognitive task (Canning et al. 2006 **A**; Paul et al. 2005). The cognitive task consisted of matching the word 'red' with the response 'yes', or the word 'blue' with the response 'no' (Bowen et al. 2001 **A**).

The overground walking component aimed to reinforce improvements in walking pattern and speed achieved on the treadmill. To reinforce the increased step length, visual cues were used in the form of non-slip footprints, which were laid at intervals normal for that subject's height. As step length approximated normal, subjects were encouraged to walk faster and were timed for feedback. Step width was reduced and balance challenged by forcing subjects to walk within one floor tile or walk along a line forwards, sideways and backwards. Workload was increased by introducing stairs and slopes to overground walking practice, and automaticity was promoted by the introduction of dual tasks. Subjects walked continuously around an outdoor circuit, which included curbs, slopes, stairs and rough terrain, while conversing with the trainer.

The immediate improvement in walking capacity measured by the 6-min. Walk Test was greater in the Ada and colleagues study than in Macko and colleagues. As described above there were differences in the programmes, which may account for these results. Macko and colleagues only used treadmill training, with increasing speed and session duration, whereas Ada and colleagues' programme involved treadmill and overground walking, focusing not only on fitness but also on quality and automaticity of walking. There is other evidence to suggest that the content of a treadmill walking programme is important in determining effectiveness. For example, Pohl and colleagues (2002 **A**) have shown the importance of manipulating the speed of the treadmill to achieve increases in overground walking speed. However, it has been shown that stroke patients generally achieve higher walking velocities by increasing their cadence rather than step length (Wagenaar et al. 1992 **A**). We therefore suggest that treadmill training programmes should include overground walking components where increases in walking speed and step length are encouraged. The improvements in walking capacity were not maintained in the Ada and colleagues study, which suggests that the one month duration was insufficient and that treadmill programmes should be of longer duration, such as the six months used by Macko and colleagues.

Based on the strategies implemented by Macko and colleagues and Ada and colleagues, we would recommend a treadmill and overground programme of 30 to 40 minutes three times a week for four to six months, with training aimed to increase speed, step length, aerobic intensity and automaticity.

PLAN: TO IMPROVE MR IB'S COMMUNITY AMBULATION

A specific intervention plan for Mr IB, based on the above evidence, is outlined below:

Gain medical clearance and consent to participate in exercise programme

Consult with Mr IB's treating doctor to organise medical clearance or stress test (as per ASCM guidelines) to participate in a treadmill and overground walking programme aimed at improving walking capacity and aerobic fitness.

Clinical facility: supervised treadmill overground walking programme focused on improving step length

Arrange for Mr IB to attend ambulatory care/out-patient setting three times a week for two weeks. Negotiate with his daughter to provide transport or organise community transport. At the initial session, determine if other impairments are interfering with walking and if so recommend treatment or adaptation. Mr IB's impaired sensation may be a reason he cannot walk backwards, since in this situation he has no peripheral vision of his feet. Commence a supervised treadmill and overground walking programme focusing on increasing step length and then increasing speed and step length.

Include a warm up, involving stretches of the calf and hip flexor muscles against a wall, before Mr IB gets on treadmill.

Local gym: treadmill training focusing on aerobic training

Consult with Mr IB and his daughter to find a local gym with a treadmill which Mr IB can access without relying on her for transport. Use a heart rate monitor (pulse monitor on treadmill) and aim to build up to training at 60–70 % of heart rate reserve for 30–40 minutes, three days a week for 8–12 weeks. Involve a personal trainer if Mr IB can afford it, and have the trainer monitor frequency, intensity and duration as well as encourage long steps. Therapist to call Mr IB one to two days a week to monitor the programme and enhance compliance.

Clinical facility: supervised treadmill and overground programme focused on automaticity

Arrange for Mr IB's daughter to bring him into the ambulatory care setting three days a week for two weeks. In these sessions, work on automaticity by introducing dual tasks, both on the treadmill and on an outdoor circuit with slopes, curbs and gutters.

Home visit

Devise a maintenance programme which Mr IB is contracted to complete. It should involve walking in his own community, for example, to the shops, around the block, accessing public transport. It may include continued attendance at the gym.

Monitoring progression and enhancing compliance

Organise regular weekly phone calls to discuss and progress monitoring programme. Institute formal reviews either in the community or at the facility to measure his walking using the 6 min. Walk Test every one to two months and progress his programme accordingly. In addition, a maintenance programme needs to be instituted and regularly reviewed to ensure that gains in walking capacity and improvements in physical activity are maintained over the long term.

CONCLUSION

In this chapter we have presented two cases in which treadmill training has been considered as an intervention to improve walking after stroke. We have highlighted the fact that the challenge for clinicians is to determine the most appropriate intervention in light of current high level evidence (systematic reviews, randomised controlled trials), weaker evidence (uncontrolled trials), observational studies, clinical experience and common sense. We argue that, while there is no conclusive high level evidence that treadmill training is effective, for other reasons treadmill training is worthy of implementation, and have given practical advice about how to implement treadmill training, both to establish walking in a non-ambulatory patient and to improve ambulation in a person residing in the community. The intervention plans reflect a balance between current evidence, clinical experience and common sense. It is essential they are regularly reviewed and updated as new evidence comes to light.

ACKNOWLEDGEMENTS

We would like to acknowledge the contribution of the clinicians, particularly Stephanie Potts and Ohnmar Aung, who are helping us undertake the randomised trial of the effectiveness of using treadmill and BWS in establishing walking in non-ambulatory patients after stroke. We thank them for sharing their experiences of implementing this intervention.

REFERENCES

Ada L, Dean CM, Hall JM, Bampton J, Crompton S (2003) A treadmill and overground walking program improves walking in individuals residing in the community after stroke: a placebo-controlled, randomized trial. *Archives of Physical Medicine and Rehabilitation* **84**(10): 1486–1491.

Bowen A, Wenman R, Mickelborough J, Foster J, Hill E, Tallis R (2001) Dual-task effects of talking while walking on velocity and balance following stroke. *Age and Ageing* **30**: 319–323.

Chen G, Patten C, Kothari DH, Zajac FE (2005a) Gait differences between individuals with post-stroke hemiparesis and non-disabled controls at matched speeds: *Gait and Posture* **22**(1): 51–56.

Chen G, Patten C, Kothari DH, Zajac FE (2005b) Gait deviations associated with post-stroke hemiparesis: improvement during treadmill walking using weight support, speed, support stiffness, and handrail hold. *Gait and Posture* **22**(1): 57–62.

Canning C, Ada L, Paul SS (2006) Is automaticity of walking regained after stroke? *Disability and Rehabilitation* **28**(2): 97–102.

Carr JH, Shepherd RB, Nordholm L, Lynne D (1985) Investigation of a new motor assessment scale for stroke patients. *Physical Therapy* **65**: 175–180.

da Cunha Filho IT, Lim PA, Qureshy H, Henson H, Monga T, Protas EJ (2002) Gait outcomes after acute stroke rehabilitation with supported treadmill ambulation training: a randomized controlled pilot study. *Archives of Physical Medicine and Rehabilitation* **83**(9): 1258–1265.

Dean CM, Richard CL, Malouin F (2001) Walking speed over 10 metres overestimates locomotor capacity after stroke. *Clinical Rehabilitation* **15**(4): 415–21.

Enright PL, Sherrill D (1998) Reference equations for the six-minute walk in healthy adults. *American Journal of Respiratory and Critical Care Medicine* **158**(5 Pt 1): 1384–1387.

Hassid E, Rose D, Commisaro J, Guttry M, Dobkin BH (1997) Improved gait symmetry in hemiparetic stroke patients induced during body weight-supported treadmill stepping. *Journal of Neurologic Rehabilitation* **11**: 21–26.

Hesse S, Helm B, Krajnik J, Gregoric M, Mauritz K-H (1997) Treadmill training with partial body weight support: influence of body weight release on the gait of hemiparetic patients. *Journal of Neurologic Rehabilitation* **11**:15–20.

Hesse S, Konrad M, Uhlenbrock D (1999) Treadmill walking with partial body weight support versus floor walking in hemiparetic subjects. *Archives of Physical Medicine and Rehabilitation* **80**: 421–427.

Kosak MC, Reding MJ (2000) Comparison of partial body weight-supported treadmill gait training versus aggressive bracing assisted walking post stroke. *Neurorehabilitation and Neural Repair* **14**: 13–19.

Macko RF, Ivey FM, Forrester LW, Hanley D, Sorkin JD, Katzel LI et al. (2005) Treadmill exercise rehabilitation improves ambulatory function and cardiovascular fitness in patients with chronic stroke: a randomized, controlled trial. *Stroke* **36**(10): 2206–2211.

Macko RF, Smith GV, Dobrovoiny CL, Sorkin JL, Goldberg AP, Silver KH (2001) Treadmill training improves fitness reserve in chronic stroke patients. *Archives of Physical Medicine and Rehabilitation* **82**: 879–884.

Macko RF, DeSouza CA, Tretter LD, Silver KH, Smith GV, Anderson PA et al. (1997) Treadmill aerobic exercise training reduces the energy expenditure and cardiovascular demands of hemiparetic gait in chronic stroke patients. *Stroke* **28**: 326–330.

Moseley A, Stark A, Cameron I, Pollock A (2005) Treadmill training and body weight support for walking after stroke: a systematic review. *Cochrane Library* **4** http://www.thecochranelibrary.com.

Nilsson L, Carlsson J, Danielsson A, Fugl-Meyer A, Hellstrom K, Kristensen L et al. (2001) Walking training of patients with hemiparesis at an early stage after stroke: a comparison of walking training on a treadmill with body weight support and walking training on the ground. *Clinical Rehabilitation* **15**: 515–527.

Paul SS, Ada L, Canning C (2005) Automaticity of walking – implications for physiotherapy practice. *Physical Therapy Reviews* **10**: 15–23.

Pohl M, Mehrholz J, Ritschel C, Ruckriern S (2002) Speed-dependent treadmill training in ambulatory hemiparetic stroke patients: a randomized controlled trial. *Stroke* **33**: 553–558.

Regensteiner JG, Steiner JF, Panzer RI (1990) Evaluation of walking impairment by questionnaire in patients with peripheral arterial disease. *Journal of Vascular Medicine and Biology* **2**: 142–152.

Scheidtmann K, Brunner H, Muller F, Weinandy-Trapp M, Wulf D, Koenig E (1999) Treadmill training in early poststroke patients – do timing and walking ability matter? (Sequenzefekte in der laufbandtherapie). *Neurological Rehabilitation* **5**(4): 198–202.

Silver KH, Macko RF, Forrester LW, Goldberg AP, Smith GV (2000) Effects of aerobic treadmill training on gait velocity, cadence, and gait symmetry in chronic hemiparetic stroke: a preliminary report. *Neurorehabilitation and Neural Repair* **14**: 65–71.

Smith GV, Macko RF, Silver KH, Goldberg AP (1998) Treadmill aerobic exercise improves quadriceps strength in patients with chronic hemiparesis following stroke: a preliminary report. *Journal of Neurological Rehabilitation* **12**: 111–117.

Smith GV, Silver KH, Goldberg AP, Macko RF (1999) 'Task oriented' exercise improves hamstring length and spastic reflexes in chronic stroke patients. *Stroke* **30**: 2112–2118.

Sullivan KJ, Knowlton BJ, Dobkin BH (2002) Step training with body weight support: effect of treadmill speed on practice paradigms on poststroke locomotor recovery. *Archives of Physical Medicine and Rehabilitation* **83**: 683–691.

van Peppen RP, Kwakkel G, Wood-Dauphinee S, Hendriks HJ, van der Wees PJ, Dekker J (2004a) The impact of physical therapy on functional outcomes after stroke: what's the evidence? *Clinical Rehabilitation* **18**(8): 833–862.

van Peppen RPS, der Harmeling-van Wel BC, Kollen BJ, Hobbelen JSM, Buurke JH, Halfens J et al. (2004b) Effects of physical therapy interventions in stroke patients: a systematic review (Dutch). *Nederlands Tijdschrift Voor Fysiotherapie* **114**(5):126–48.

Wade DT (1992) *Measurement in Neurological Rehabilitation* Oxford: Oxford University Press.

Wagenaar RC, Beek WJ (1992) Hemiplegic gait; a kinematic analysis using walking speed as a basis. *Journal of Biomechanics* **25**:1007.

Werner C, von Frankenberg S, Treig T, Konrad M, Hesse MD (2002) Treadmill training with partial body weight support and an electromechanical gait trainer for restoration of gait in subacute stroke patients. *Stroke* **33**: 2895–2901.

6 Treatment of the Upper Limb Following Stroke: A Critical Evaluation of Constraint Induced Movement Therapy

MARTINE NADLER

BACKGROUND

In this chapter I am going to consider the role of constraint induced movement therapy in the treatment of Mr BB, a 46 year old, right handed furniture salesman who suffered a stroke two years ago. Prior to his stroke, he was fully independent and was a keen badminton player, night-clubber and salsa dancer. He lived alone in a first floor flat and although he had little family leaving nearby, he had a circle of close friends.

DIAGNOSIS

Mr BB presented in dramatic fashion, suffering a sudden onset left hemiplegia. Investigation showed that this stroke was caused by a large right hemisphere cortical haemorrhage from the rupture of an arterio-venous malformation (AVM). The AVM was treated by surgical clipping. His symptoms were so severe that he remained in a specialist neuroscience centre for over six months and then needed six months of out-patient physiotherapy. In the early stages, when he was sufficiently medically stable to tolerate therapy, he had no sitting balance and pushed to the left. In addition to the marked physical impairments which proved such a challenge to therapy, he also had neglect of the left side. However, the paralysis was thought to be, and was treated as, the dominant feature. The arm was included in physiotherapy treatment, but at that stage it had little measurable effect.

Mr BB was discharged after a year. At his best, he was walking independently without aids. However, he needed an ankle foot orthosis for a persisting left foot-drop. He was unable to use his left hand at all and it hung limply by his side while walking.

Recent Advances in Physiotherapy. Edited by C. Partridge
© 2007 John Wiley & Sons, Ltd

SUBJECTIVE REPORT TWO YEARS AFTER STROKE

Over the year following discharge (and second year following stroke) Mr BB returned to full-time work. He adapted his work tasks to allow them to be accomplished using just the right hand. However, he could do no tasks (for example heavy lifting) that required both hands. He also adapted his lifestyle and although totally reliant on his right hand, he was fully independent. With the help of a specialist mobility centre, he even learnt to drive an automatic car, with a steering wheel knob adaptation enabling him to steer with one hand.

Mr BB was a highly motivated individual and keen to improve the use of his stroke affected arm. He was fed up of having to rely on only one arm. He considered his left arm a 'useless limp object' and worried that it adversely affected his appearance and hindered his dancing. His goal was for his arm to look more normal when walking and dancing and to have some useful function back.

OBJECTIVE EXAMINATION TWO YEARS AFTER STROKE

Mr BB's clinical picture was highly unusual. All the main muscles of his left stroke arm and shoulder were severely atrophied and the arm hung limply by his side, but in spite of this he was able to produce excellent selective movements of the fingers and thumb. For example, he could rapidly tap his thumb to each of his fingers in turn. The identification of these fractionated finger movements was very important because it indicated that there was significant corticospinal tract innervation to these muscles. The corticospinal tract is the most important motor tract, connecting the motor cortex via the anterior horn cells in the spinal cord to the peripheral muscles. It is the only tract that enables fine finger movements to be carried out. If the corticospinal tract is still innervating as far distally as the fingers it is very likely that more proximal innervation of the arm muscles is present, even if not used. Our hypothesis was that the dyspraxia that Mr BB had exhibited from the start was now the major factor restricting the use of his left arm. He should theoretically be able to activate the proximal muscles of the upper limb.

For this discussion I shall define dyspraxia as the inability to execute previously learnt motor patterns.

SUMMARY

Following recovery from the stroke, Mr BB was walking independently and wore an ankle foot orthosis (AFO) on the left leg. His left upper limb had no useful activity. There was atrophy visible in all the muscle groups. Weakness was demonstrable in all groups but sensation was normal throughout. There was active wrist extension to the neutral position. He had a full active range of selective finger extension and flexion, and selective grasp and release.

ASSESSMENT AND OUTCOME MEASUREMENTS

MEASUREMENT OF FUNCTIONAL UPPER LIMB ACTIVITY

There are many assessment tools available. Some of the more popular in clinical practice are the Motor Assessment Scale (Carr et al. 1985), Action Research Arm Test (Lyle 1981) and Motricity Index (Demeurisse et al. 1980). A comprehensive overview of these and other measurement tools is contained in Wade (1992).

MEASUREMENT OF MUSCLE WEAKNESS

The muscle weakness following stroke can be graded using the MRC Oxford Scale of muscle strength, or measured using a hand-held myometer. This instrument measures the maximum isometric muscle strength in a standardised position (Bohannon 1989).

MEASUREMENT OF JOINT RANGE

Active and passive ranges of upper limb joint movement should be measured with a goniometer.

MEASUREMENT OF LOWER LIMB FUNCTION

Having a non-functioning arm may impact on the quality and speed of walking. This is because the stroke arm acts as a dead weight, dragging on and changing the alignment of the trunk and making it more difficult to balance on the stroke leg. Therefore it is important to measure the walking ability. The self-paced 10 metre timed walk (Bradstater et al. 1983) is a good tool for this purpose.

GENERAL PRINCIPLES OF TREATMENT

Mr BB had clearly adapted extremely well to using his sound right upper limb to compensate for the deficits in the left stroke hand. However, examination revealed that he had some recovery of the left stroke hand but failed to utilise this potential. Taub and colleagues have hypothesised that a proportion of the motor deficits in the upper limb which persist after stroke may result from learned behaviour, which they call 'learned non-use'. The process may be summarised as follows. In the initial stages after a stroke, the patient is unable to use the stroke affected upper limb due to the neural damage. If the patient finds use of the stroke hand futile, he adapts and learns not to try to use it. Instead he learns to compensate, relying on the healthy hand to function. Later, there may be some recovery in the stroke hand but by then, the patient has learned not to use it. Thus, recovery is masked by 'learned non-use' (Taub et al. 1993 **A**; Taub et al. 2002 **R**).

Constraint induced movement therapy (CIMT) is a therapeutic approach which combines intensive training of the stroke affected upper limb with the wearing of a restraint (for example a sling or a mitten) on the non-stroke arm. The rationale is that by combining these two elements (intensive practice and restraint), the learned non-use may be reversed and the full potential of the upper limb function realised, with sustained functional improvement.

WHAT DOES CONSTRAINT INDUCED MOVEMENT THERAPY INVOLVE?

Traditionally, CIMT involves six to seven hours of behavioural shaping therapy, with a therapist providing individual input to a patient, every weekday for two consecutive weeks. Out of therapy, the patient wears a restraint on the sound side for 90 % of their waking hours.

Behavioural shaping is an approach developed from the field of neuropsychology and is:

> ... a training method in which a desired motor or behavioural objective is approached in small steps by "successive approximations" so that the amount of improvement required for successful performance at each step is always small. Taub & Wolf 1997 **R**.

This intensive CIMT input has been shown to provide lasting improvement in stroke upper limb function up to two years post-study (Kunkel et al. 1999 **A**; Miltner et al. 1999 **A**; Taub et al. 1993 **A**). This was measured using the Wolf Motor Function Test, which measures limb movement, and the Motor Activity Log (including Actual Amount of Use Test and Quality of Movement), which measures how much the patient uses their stroke limb for a series of tasks in 'real life' during the day. Delivering the same quantity of CIMT over a longer time frame (for example, three hours of behavioural shaping during weekdays over four weeks) showed similar functional improvement (Dettmers et al. 2005 **A**). Sterr et al. (2002 **A**) have tested patients undergoing three hours of behavioural shaping per day, compared to six hours per day, for a fortnight. In their small randomised controlled trial ($n = 15$), both groups showed significantly improved arm function, but effects were greater in the group who underwent six hours daily than in those who underwent three hours daily (Sterr et al. 2002 **A**).

Behavioural shaping is not part of a typical physiotherapy treatment repertoire, although physiotherapists may informally use similar principles. For example, they adapt their treatment so that the patient practises activities which are achievable with a little assistance. Practising tasks which are too easy is unlikely to promote motor learning or improve function and, conversely, very difficult tasks fail to improve function due to lack of motivation.

The CIMT protocol as outlined by Taub's group is a costly use of resources, with one therapist treating an individual patient for six to seven hours a day. In order to identify how practical this approach would be, a survey was carried out in the USA. Results showed that 68 % of patients with stroke reported that they would not like to participate in the standard CIMT protocol. Of those who would participate, two thirds said that they were unlikely to adhere to the protocol. In addition, over 60 % of therapists surveyed felt that non-compliance would be a problem, and the majority felt that there was a lack of resources or facilities (Page et al. 2002 **A**).

QUESTION 1

Is a constraint induced movement therapy approach appropriate for Mr BB?

The inclusion criteria are that patients can actively produce 20 degrees of wrist extension and 10 degrees of finger extension (stroke side), can walk safely without a walking aid, lack cognitive impairment, and are more than a year post-stroke. Mr BB fits these inclusion criteria.

However, Mr BB was unable to have CIMT delivered according to this strict protocol because the physiotherapist did not have the specialist training in behavioural shaping and there were inadequate resources to deliver this intensity of treatment in the current NHS climate.

QUESTION 2

Can a modified constraint induced movement therapy be used to improve stroke upper limb function?

A very valuable lesson from CIMT is the importance of repeated practice and intensive use of the stroke upper limb. The CIMT programme may be modified in four ways. Firstly, rather than strictly adhering to behavioural shaping principles, practice of specific activities of daily living or components of these using his stroke hand could be included. Secondly, treatment could be given in a group setting. Thirdly, time spent in group therapy could be reduced, with the patient undertaking to practice specific tasks set for him in his own time. Fourthly, restraint could be used during therapy time alone.

The evidence for modified CIMT comes from a number of studies. The largest was a randomised controlled trial of 66 patients (Van der Lee et al. 1999 **A**). In this study, all patients were treated in groups of four, supervised by one to two therapists per group. The experimental group of patients received forced use treatment (ADL type activities) for two weeks (six hours per day) and wore a restraint on the non-stroke side in therapy sessions, keeping a log of how much it was worn during waking hours. The control group had equally intensive input (without restraint), which comprised bi-manual training for the same time period using the neurodevelopmental technique. One week after the intervention, results showed small but significant improvement in the experimental group compared to the control group for the Action Research Arm Test (dexterity measure) and for the Motor Activity Log (Actual Amount of Use

Test) post-treatment. However, at one year follow-up, significant differences were only apparent in the Action Research Arm Test. On closer analysis, the subgroups that benefited most from forced use were patients with hemi-neglect and sensory problems (Van der Lee et al. 1999 A).

This amount of input may still be difficult to deliver with current resources. A small study ($n = 6$) tested a different modified form of CIMT. 30 minutes of physiotherapy combined with 30 minutes of occupational therapy three times a week for 10 weeks, combined with five hours of task practice with restraint a day at home, produced functional improvement on the Fugl-Meyer, Wolf Motor Function Test and the Action Research Arm Test (Page et al. 2001 A). Improvements were not noted in the patients who underwent conventional treatment or had no treatment at all. In another study, Johansen-Berg et al. (2002 A) gave a small group of chronic stroke patients ($n = 7$) a 30 minute programme of graded exercises to be carried out twice daily for two weeks while wearing a restraint on the healthy upper limb; either a sling or a mitten depending on whether the healthy arm was needed for balance. Results showed improvements in grip strength in the affected hand (Johansen-Berg et al. 2002 A).

Therefore, I would explore the possibility of treating Mr BB with a modified form of CIMT, depending on local resources. I would encourage him to take leave from work and treat him in a group setting daily for two weeks. In this group setting, activities could be carried out using a circuit format, where participants spend 10 minutes on each task before rotating to the next one, as suggested by van der Lee (1999 A). Examples could include ADL type activities, such as hanging clothes, opening pegs, opening jars or tupperwares, and cutting fruit or vegetables. At the same time, I would recommend that Mr BB wear a mitten on his healthy hand to discourage its use and focus attention on learning to reuse the stroke hand. Given the muscle weakness and atrophy, tasks might initially need to be carried out with gravity neutralised before progressing to exercises against gravity.

As his treating physiotherapist, I would recommend Mr BB undertake five hours of daily practice at home to reinforce the use of the stroke affected upper limb. In order to maintain motivation and to access previous motor patterns I would discuss goals and tailor treatment accordingly, taking into account his occupation and leisure interests. I would not recommend his wearing a restraint on his healthy hand, both for safety reasons and because many functional tasks require the use of both hands.

Thus, for Mr BB task practice might include holding a tape measure with both hands. He could practise throwing a shuttlecock with his stroke hand to serve with his right hand. He could start dancing with a partner using both hands. Given his plans to return to studying, practising using a computer keyboard with both hands would be useful. He could also try to hold the steering wheel of the car with his affected hand while using his healthy hand to steer. Texting messages on his mobile phone would recruit and refine thumb activity. The use of visual markers (for example, a red dot on his glass, toothbrush, tap, shower control) could serve as a cue reminding him to use his left stroke hand.

QUESTION 3

Is there any evidence to suggest that neurophysiological changes accompany clinical improvements?

A number of studies have used transcranial magnetic stimulation (TMS), delivered to the motor cortex before and after chronic stroke patients underwent CIMT, using the Taub et al. (1993 **A**) protocol. Liepert et al. (1998 **A**) have shown that, after two weeks of CIMT, the number of active cortical sites and the area representation of the abductor pollicis brevis thumb muscle were increased and shifted on the stroke-affected hemisphere. These cortical map changes were shown to accompany functional improvement (Liepert et al. 2000 **A**) and occurred after CIMT was carried out in addition to, rather than after, conventional therapy alone (Liepert et al. 2001 **A**). The authors hypothesised these cortical representation changes were due to increased cortical excitability. This may result from decreased activity of local inhibitory interneurones, unmasking of existing synaptic connections, and/or increased strength of existing connections. These findings are supported by Wittenberg et al. (2003 **A**), who used positron emission tomography to show more normal activation of the affected primary sensorimotor cortex during movement of the affected hand, which they hypothesised was due to more efficient recruitment of neurons. Using functional magnetic resonance imaging, increased activation of the damaged pre-motor cortex correlated with improved grip strength of the paretic hand (Johansen-Berg et al. 2002 **A**). It is unclear whether the changes were due to wearing a restraint or to the intensive practice.

CRITICAL EVALUATION OF THE EVIDENCE

There is some evidence to suggest that CIMT or modified CIMT may be beneficial in the rehabilitation of upper limb function following stroke. However, the most dramatic changes have been reported in studies which are uncontrolled single or multiple case series (Dettmers et al. 2005A; Kunkel et al. 1999 **A**; Miltner et al. 1999 **A**), rather than in a randomised controlled trial. This may exaggerate the treatment effect and fail to compare CIMT intervention with a control. In a review, van der Lee (2001 **R**) considered that the evidence for the effectiveness of CIMT was somewhat limited and concluded that it was simply the intensity of treatment delivered which was responsible for the functional improvement, rather than the use of a restraint. The author concluded that CIMT may not be a different treatment as such but simply 'more of the same'.

REFERENCES

Bohannon (1989) Correlation of lower limb strengths and other variables with standing performance in patients with brain lesions. *Physiotherapy Canada* **41**: 198–202.

Bradstater ME, de Bruin H, Gowland C, Clarke BM (1983) Hemiplegic gait: analysis of temporal variables. *Archives of Physical Medicine and Rehabilitation* **64**: 583–587.

Carr JH, Shepherd RB, Nordholm L, Lynn D (1985) Investigation of a new motor assessment scale for stroke patients. *Physical Therapy* **65**: 175–180.

Demeurisse G, Demol O, Robaye E (1980) Motor evaluation in vascular hemiplegia. *European Neurology* **19**: 382–9.

Dettmers C, Teske U, Hamzei F, Uswatte G, Taub E (2005) Distributed form of constraint induced movement therapy improves functional outcome and quality of life. *Archives of Physical and Medical Rehabilitation* **86**: 204–209.

Johansen-Berg H, Dawes H, Guy C, Smith SM, Wade DT, Matthews PM (2002) Correlation between motor improvements and altered fMRI activity after rehabilitative therapy. *Brain* **125**: 2371–2742.

Kunkel A, Kopp B, Muller G, Villringer K, Villringer A, Taub E et al. (1999) Constraint induced movement therapy for motor recovery in chronic stroke patients. *Archives of Physical and Medical Rehabilitation* **80**: 624–628.

Liepert J, Bauder H, Miltner WHR, Taub E, Weiller C (2000) Treatment-induced cortical reorganization after stroke in humans. *Stroke* **31**(6): 1216.

Liepert J, Miltner WHR, Bauder H, Sommer M, Dettmers C, Taub E et al. (1998) Motor cortex plasticity during constraint-induced movement therapy in stroke patients. *Neuroscience Letters* **250**: 5–8.

Liepert J, Uhde I, Graf S, Leidner O, Weiller C (2001) Motor cortex plasticity during forced-use therapy in stroke patients: a preliminary study. *Journal of Neurology* **248**: 315–321.

Lyle RC (1981) A performance for assessment of upper limb function in physical rehabilitation and research. *International Journal of Rehabilitation Research* **4**: 483–493.

Miltner HR, Bauder H, Sommer M, Dettmers C, Taub E (1999) Effects of constraint induced movement therapy on patients with chronic motor deficits after stroke: a replication. *Stroke* **30**: 586–592.

Page SJ, Levine P, Sisto S, Bond Q, Johnston MV (2002) Stroke patients' and therapists' opinions of constraint induced movement therapy. *Clinical Rehabilitation* **16**: 55–60.

Page SJ, Sisto S, Levine P, Johnston MV, Hughes M (2001) Modified constraint induced therapy: a randomized feasibility and efficacy study. *Journal of Rehabilitation Research and Development* **38**: 583–590.

Sterr A, Elbert T, Berthold I, Kolbel S, Rockstroh B, Taub E (2002) Longer versus shorter daily constraint induced movement therapy of chronic hemiparesis: an exploratory study. *Archives of Physical and Medical Rehabilitation* **83**: 1374–1377.

Taub E, Miller NE, Novack TA, Cook III EW, Fleming WC, Nepomuenco CS et al. (1993) Technique to improve chronic motor deficit after stroke. *Archives of Physical and Medical Rehabilitation* **74**: 347–354.

Taub E, Wolf SL (1997) Constraint induced movement techniques to facilitate upper extremity use in stroke patients. *Topics in Stroke Rehabilitation* **3**(4): 38–61.

Taub E, Uswatte G, Elbert T (2002) New treatments in neurorehabilitation founded on basic research. *Nature Reviews Neuroscience* **3**: 228–236.

van der Lee JH, Wagenaar RC, Lankhorst GJ, Vogelaar TW, Deville WL, Bouter LM (1999) Forced use of the upper extremity in chronic stroke patients: results from a single-blind randomised clinical trial. *Stroke* **30**: 2369–2375.

van der Lee JH (2001) Constraint induced therapy for stroke: more of the same or something completely different? *Current Opinion in Neurology* **14**: 741–744.

Wade D (1992) *Measurement in Neurological Rehabilitation*. Oxford: Oxford University Press.

Wittenberg GF, Chen R, Ishii K, Bushara KO, Taub E, Gerber LH et al. (2003) Constraint induced therapy in stroke: magnetic stimulation motor maps and cerebral activation. *Neurorehabilitation and Neural Repair* **17**(1): 48–57.

IV Pain Management

7.1 An Introduction to Current Concepts of Pain

LESTER JONES

PAIN: A DEFINITION

Human pain is complex. It is a multi-dimensional subjective experience that can be described as a perceptual response to all types of stimuli that threaten the person's homeostasis (Gifford 1998 **C**; Moseley 2003 **C**; Henderson et al. 2005 **R**). While pain has also been described as a 'multiple system output' (Moseley 2003 **C**, p. 130), the definition of pain that is presented here was developed by the International Association for the Study of Pain (IASP). It states that: 'Pain is an unpleasant sensory and emotional experience associated with actual or potential tissue damage, or described in terms of such damage' (Merskey & Bogduk 1994 **C**, p. 210).

It will be valuable, for Chapters 7.1–7.3 focusing on pain, to consider this definition in detail.

SENSORY COMPONENT

The first point to consider is quite unexciting: sensory processes are involved in the perception of pain. This is nothing new. However, it is worth highlighting the term nociception. Nociception describes the recognition of noxious[1] stimuli by specific sensory receptors (for example, nociceptors) and in turn the transmission of nerve impulses to the central nervous system (for reviews, see Basbaum et al. 2005 **R**; Galea 2002 **R**). That is, it is a sensory physiological process that could be interpreted as the sensory component of pain.

The second, more interesting point is that pain has a sensory component, that is, it is not entirely sensory. Importantly then, nociception is not pain. This challenges the traditional emphasis on tissue damage, inflammation processes and disease processes in explaining pain.

---

[1] Noxious stimuli are stimuli that are causing, or potentially could cause, tissue damage.

Recent Advances in Physiotherapy. **Edited by C. Partridge**
© 2007 John Wiley & Sons, Ltd

EMOTIONAL COMPONENT

According to the definition, there is also an emotional component to pain. It is easy to establish a relationship between emotion and pain. Anyone who has stubbed a toe or jammed a finger will recall the anger, distress or fear that was associated with their pain perception – perhaps, they will reflect, disproportionately so.

The definition does not entail a mere relationship between pain and emotions, but that pain actually has an emotional component. That is, whenever someone feels pain, their emotional state is playing a part (Klaber Moffett 2000 **C**; Price 2000 **R**). Linton (2005 **A**) identifies distress and pain catastrophising[2] as strong predictors of onset of back pain, possibly mediated by anxiety. This recognises the interdependence of cognitive and emotional factors, and suggests it may be more accurate to consider a cognitive-emotional component, rather than simply an emotional one. This is reinforced by evidence that education can have an effect on anxiety and post-operative pain (Carr & Goudas 1998 **R**). A person's perception of their own pain is therefore influenced by both a sensory component and an emotional-cognitive component, and physiotherapists need to strive to understand and manage both.

It may be helpful in understanding the cognitive-emotional component to consider that depression, and other negative emotional states, can lead to a person feeling globally vulnerable. As a result, the processing of all types of potentially threatening stimulation detected by the various receptors of the body is prioritised. The nervous system becomes hyperresponsive or hypervigilant;[3] nociceptive processes become sensitised, with an increased responsiveness to non-noxious sensory stimuli and reduced activation thresholds at nociceptors (Flor et al. 2004 **A**; Mitchell et al. 2000 **R**; Villemure & Bushnell 2002 **R**). This enables low level stimuli to create activity in the nociceptive system (for example, touch can cause pain). Links between depression and pain (Williams et al. 2006 **R**) and anxiety and pain (Linton 2005 **R**) seem well established. It could be that the person who perceives themselves as vulnerable, is predisposed to pain. In contrast, if the person can be made to feel less vulnerable, then the state of the nervous system will be normalised, and the likelihood of feeling pain may be reduced.

A more focused increase in nervous system responsiveness may apply when particular parts of the body are perceived as vulnerable. Fear of damage, re-injury or increased pain may provide the emotional stimulus here, leading to the belief that a particular part of the body is under threat. All incoming information from intero and exteroceptors located in, or relating to, the body parts that are perceived as vulnerable would therefore be potentially threatening. This could result in more attention from the central nervous system, due to the need for action: protection or escape (Crombez

---

[2] Pain catastrophising can be defined as a response to pain in which a person dwells on, or magnifies the potential for, the negative consequences of their pain. It may include statements about inability to cope with pain.

[3] Hypervigilance can be considered to be a partly automatic response, where the brain attends to information relating to a threat or fear, regardless of (and potentially competing with) the task the person is occupied with (Crombez et al. 2005 **R**).

et al. 2005 **R**). The combination of perceived threat and perceived need to act may be fundamental to a person's perception of their pain (Moseley 2003 **C**). As such, consideration of emotional factors, and the associated cognitions relating to vulnerability, is likely to be important in the assessment and treatment of pain. Recognising that pain isn't just associated with emotions but is in part emotional, is not only the first step in accepting the IASP definition, but the first step in understanding the complexities of human pain.

PAIN IS AN EXPERIENCE

Pain is described as a 'sensory and emotional experience' (Merskey & Bogduk 1994, p. 210). The use of the word 'experience' reinforces the perceptual nature of pain and identifies it as personal and therefore individual. This reflects the fact that pain – like other perceptions – is influenced by current context, past experiences, and expectations, including motor planning (Schuchert 2004 **C**). In evaluating attention and learning, both linked to the pain experience, Schuchert suggests that 'motor planning is in effect before the processing of a stimulus is complete, such that the anticipation of an action response actually assists and shapes the processing of a stimulus' (p. 160). If this widely held view is true, patients may demonstrate more pain behaviour when they are engaged in a consultation about their pain, or when attempting activity they perceive as pain-provoking. When they are doing something away from that context, their pain and related behaviour lessens because the anticipation of pain is reduced. In the past, this mismatch of behaviour may have been interpreted as malingering. Hopefully physiotherapists no longer make this reasoning error, but recognise that a person's pain experience can vary in different environments and contexts.

The physiotherapist also needs to remember that when a patient reports pain they may not necessarily be able to, or willing to, describe their pain experience (Bendelow 2000 **C**; Keefe et al. 2000 **A**; Williams et al. 2000 **A**). A person's report of pain is only an indication of their sensory and emotional experience, and reflects cognitive factors such as beliefs about pain and perceived threat, as well as communication abilities. The accuracy of the description of the pain experience is also limited by the accuracy of the person's internal model of their own body – the so-called body schema or virtual body (Moseley 2003a **C**) held within the brain. The virtual body is susceptible to distortions; for example, phantom limbs in amputees. Despite the potential for inaccuracies, the report of pain is often the only reasonable indicator that is accessible when making health care management decisions. As such, this subjective information needs careful evaluation before it is used to drive treatment planning.

PAIN AND TISSUE DAMAGE

It is common to relate pain to tissue damage. The IASP definition incorporates this well-held belief but adds that pain does not require actual tissue damage, but may simply be associated with a description of tissue damage. The somewhat controversial point that can be drawn from this is that pain can exist even when there is no evidence of tissue damage. The logical conclusion is to suggest a psychological origin to the pain (psychogenic), which does not involve the sensory system.

However, this must be qualified. First, despite well developed strategies for identifying tissue pathology, there is no guarantee that investigations can target all potential sensory triggers. This was one of the conclusions of a review looking into the cause of tendon pain (Khan et al. 1999 **R**). Second, damage to neural structures may cause ectopic impulses that lead to a persistent input promoting centrally-mediated pain (McMahon 2002 **R**). Third, the potential for emotions and cognitions to alter the sensitivity of the nervous system appears to be extremely powerful (Benedetti et al. 2003 **A**; Graceley et al. 2004 **A**; Petrovic & Ingvar 2002 **R**; Price 2000 **R**). Therefore, it is important to consider the impact of psychological factors on the sensitivity of the nervous system – making possible the involvement of sensory stimuli not related to tissue damage – before concluding that the pain experience is being caused by psychological factors alone.

The familiar perception, itch, can be used as an example of the ability of the brain to integrate psychological and sensory components in perception. Similar to pain, it is associated with nociceptive stimuli (Magerl 1996 **C**). Ask a person if they have an itch somewhere and the person's nervous system begins scanning the inputs it is receiving (vigilance). Inevitably an itch is found. Further, if someone talks about something that causes itch (for example, mosquitoes, head lice) then the brain of the receiver of that information will become alert to this sensation and again an itch will often be detected. The sensation is not being created, it is already there. So itch would appear to be mediated by central processes.

The perceptions of itch and pain may be influenced by the ability of the brain to selectively respond to sensory information. That is, the vigilance of the nervous system, a mediator of attention (Eccleston & Crombez 2005 **C**), can fluctuate. This affects the sensitivity of the nervous system to nociceptive information. Because of this, distracting a person from their pain with other attention-demanding activities can be an effective but transient strategy for reducing pain (Eccleston & Crombez 2005 **C**; Villemure & Bushnell 2002 **R**).

Understanding of the multiple processes involved in the perception of pain is incomplete. However, there have been some multidimensional models developed (Gifford 1998 **C**; Melzack 1999 **C**). Moseley (2003 **C**), extending Melzack's neuromatrix model, emphasises the role of perceived danger on the activity of a 'pain neuromatrix' (p. 131). On this view, the pain neuromatrix, a network of cortical mechanisms and processors, can be activated in response to a perceived threat (perceived tissue damage) to produce an attention-demanding perceptual response (pain) and simultaneously prepare a motor output to reconcile the danger. Here pain is a warning sign, created by the central nervous system when the person or a body part is under threat, and not a sign of tissue damage per se. If Moseley is correct, and taking into account the potential individuality and changeability of the pain neuromatrix, then the complexity of the neurophysiology of pain becomes apparent. In any case, in situations where no tissue pathology has been identified, or where the evidence of psychological contribution is high, the patient's report of pain must not be downgraded; it is real

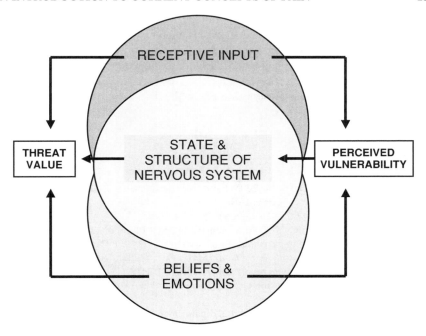

Figure 7.1.1. Influences on pain perception. Receptive input will be perceived as *threatening* due to pre-determined genetic influences on the nervous system or due to existing beliefs or emotions. As well as nociceptive input, visual input (e.g. blood; missing limb; bandage), auditory (e.g. audible cracks and clicks; being told you have a 'crumbling spine' or 'wear and tear'), proprioceptive (e.g. feelings of tightness; instability; weakness or incoordination; distorted 'virtual' body) and tactile (e.g. feeling deformity or altered temperature) input could also potentially be interpreted as threatening. Where the individual identifies a particular ('virtual') body part to be vulnerable or under threat, the nervous system may become hypervigilant to all receptive input relating to the body part, for example its sensitivity will be heightened. It is therefore suggested a specific combination of threatening receptive input and perceived vulnerability might trigger the individual pain neuromatrix with or without evidence of tissue damage. Threat value is the result of conscious and sub-conscious interpretation of input, and perceived vulnerability refers to a concept of self (whole body or part of body).

and must be legitimised (Salmon 2000 **C**). The perception of pain without evidence of tissue damage may be caused by undefined or missed tissue pathology, or by a nervous system made hypersensitive by internal beliefs and emotions.

COMPANION DEFINITION

In light of this interpretation of the IASP definition, the following statement is proposed as a companion definition:

Pain is a perception created by the brain in response to threatening receptive input (nociceptive, visual, auditory, proprioceptive, tactile) and the internal beliefs and emotions drawn from past learning. It is influenced by the structure and state of the nervous system from past and present experiences (for example, genetics, neuroplasticity, sensitivity), and associated with the priming of motor responses (preparation of action to avoid threat) and a heightened vigilance to the vulnerable part of the virtual body.

A visual representation is presented in Figure 7.1.1.

REFERENCES

Basbaum A, Bushnell MC, Devor M (2005) Pain: basic mechanisms. In: Justins DM (ed.) *Pain 2005 – An update review*. Seattle: IASP Press.

Bendelow G (2000) *Pain and Gender* London: Prentice-Hall.

Benedetti F, Pollo A, Maggi G, Vighetti S, Rainero I (2003) Placebo analgesia: from physiological mechanisms to clinical implications. In: Dostrovsky JO, Carr DB, Koltzenberg M (eds) *Proceedings of the 10th World Congress on Pain* Seattle: IASP Press.

Carr DB, Goudas LC (1998) Acute pain. *Lancet* **353**: 2051–2058.

Crombez G, Van Damme S, Eccleston C (2005). Hypervigilance to pain: an experimental and clinical analysis. *Pain* **116**: 4–7.

Eccleston C, Crombez G (2005) Attention and pain: merging behavioural and neuroscience investigations. *Pain* **113**: 7–8.

Flor H, Diers M, Birbaumer N (2004) Peripheral and electrocortical responses to painful and non-painful stimulation in chronic pain patients, tension headache patients and healthy controls. *Neuroscience Letters* **361**: 147–150.

Galea MP (2002) Neuroanatomy of the nociceptive system. In: Strong J, Unruh AM, Wright A, Baxter GD (eds) *Pain: a textbook for therapists* London: Harcourt Publishers Limited.

Gifford LS (1998) Pain, the tissues and the nervous system: a conceptual model. *Physiotherapy* **84**(1): 27–36.

Graceley RH, Geisser ME, Giesecke T, Grant MAB, Petzke F, Williams DA et al. (2004) Pain catastrophizing and neural responses to pain among persons with fibromyalgia. *Brain* **127**: 835–843.

Henderson M, Kidd BL, Pearson RM, White PD (2005) Chronic upper limb pain: an exploration of the biopsychosocial model. *Journal of Rheumatology* **32**: 118–122.

Keefe FJ, Lefebvre JC, Egert JR, Affleck G, Sullivan MJ, Caldwell DS (2000) The relationship of gender to pain, pain behaviour, and disability in osteoarthritis patients: the role of catastrophising. *Pain* **87**: 325–334.

Khan KM, Cook JL, Bonar F, Harcourt P, Astrom M (1999) Histopathology of common tendinopathies. Update and implications for clinical management. *Sports Medicine* **27**(6): 393–408.

Klaber Moffett J (2000) Pain: perception and attitudes. In: Gifford L (ed.) *Topical Issues in Pain 2* Falmouth: CNS Press.

Linton SJ (2005) Do psychological factors increase the risk for back pain in the general population in both a cross-sectional and prospective analysis? *European Journal of Pain* **9**(4): 354–361.

Magerl W (1996) Neural mechanisms of itch sensation. Technical Corner from *IASP Newsletter* http://www.iasp-pain.org/TC96SeptOct.html Accessed 26 May 2006.

McMahon SB (2002) Neuropathic mechanisms. In: Giamberardino MA (ed.) *Pain 2002 – An update review*. Seattle: IASP Press.

Melzack R (1999) From the gate to the neuromatrix. *Pain* **6** Suppl.: 121S–126S.

Merskey H, Bogduk N (1994) *Classification of Chronic Pain: descriptions of chronic pain syndromes and definition of chronic pain terms* (2 edn) Seattle: IASP Press.

Mitchell S, Cooper C, Martyn C, Coggon D (2000) Sensory neural processing in work-related upper limb disorders. *Occupational Medicine* **50**(1): 30–32.

Moseley GL (2003) A pain neuromatrix approach to patients with chronic pain. *Manual Therapy* **8**(3): 130–140.

Petrovic P, Ingvar M (2002) Imaging cognitive modulation of pain processing. *Pain* **95**: 1–5.

Price DD (2000) Psychological and neural mechanisms of the affective dimension of pain. *Science* **288**: 1769–1772.

Salmon P (2000) Patients who present physical symptoms in the absence of physical pathology: a challenge to existing models of doctor-patient interaction. *Patient Education and Counselling* **39**: 105–113.

Schuchert SA (2004) The neurobiology of attention. In: Schumann JH, Crowell SE, Jones NE, Lee N, Schuchert SA, Wood LA *The Neurobiology of Learning* London: Lawrence Erlbaum Associates, pp. 143–173.

Villemure C, Bushnell MC (2002) Cognitive modulation of pain: how do attention and emotion influence pain processing? *Pain* **95**(3): 195–196.

Williams AC de C, Oakley Davies HT, Chadury Y (2000) Simple pain rating scales hide complex idiosyncratic meanings. *Pain* **85**(3): 457–463.

Williams LJ, Jacka FN, Pasco JA, Dodd S, Berk M (2006) Depression and pain: an overview. *Acta Psychiatrica* **18**: 79–87.

7.2 Non-Specific Arm Pain

LESTER JONES

CASE REPORT

BACKGROUND

Miss NS is a 25 year old woman and lives alone. Her parents, whom she regularly visits, live four hours' drive away. She works as an administrative assistant in a busy human resources department for a large newspaper. The nature of the work has changed over the last six months, with greater emphasis on keyboarding tasks, and generally she feels the workload has increased. She feels that her manager is not always sympathetic to staff concerns about stress and workload.

She developed pain in her right elbow region that was exacerbated with note-taking and keyboarding. A work station assessment was carried out, with some modifications and advice given, but symptoms persisted and she was seen by the occupational health doctor at her workplace. He referred her to her general practitioner (GP) in order to organise physiotherapy. A private physiotherapy appointment was made four weeks after initial onset of symptoms.

The two colleagues with whom she works most closely have had similar symptoms across the previous 18 months and one had surgery in an attempt to resolve the problem.

Miss NS is considering looking for another job as a result of the workplace stress and her work-related symptoms.

MEDICAL DIAGNOSIS

She was referred to her GP by the occupational health doctor with diagnosis/label of 'tennis elbow'.

She was referred to physiotherapy by her GP with diagnosis/label of 'tendinitis' or 'RSI'.

ASSESSMENT

Initial presentation to physiotherapy

- Pain spreading proximally and distally in right arm.

Recent Advances in Physiotherapy. **Edited by C. Partridge**
© 2007 John Wiley & Sons, Ltd

- Remains independent but now hair-washing, long-distance driving, some cooking tasks affected; modifies rather than avoids activity.
- No time off work.
- NSAID no effect.
- Wears an elasticised tubular bandage on right forearm/elbow.

On examination

- Right arm resting on lap and no automatic gesturing.
- Tenderness with palpation over and around common extensor insertion and into bulk of wrist extensors.
- Increased muscle tone in right forearm, upper and middle trapezius muscles and right pectoral muscles.
- Palpation of right arm elicits a discomfort that is difficult for Miss NS to describe but is unpleasant.
- Joint movement (quality and range):
  - Right elbow – reduced speed and guarding, especially with extension; pain with extension but full range of movement (FRoM).
  - Right shoulder – reduced speed in elevation; no pain but stiffness, especially at end of range (FRoM).
  - Cervical spine – some discomfort with flexion and also with lateral flexion to left and stiffness end of range (FRoM).
- Muscle extensibility: reduced in right wrist extensors, right elbow flexors, right pectoral muscles, cervico-scapular muscles.
- Neurodynamic upper limb test: range of elbow extension is reduced (right vs left) with radial nerve bias, wrist flexed, and cervical spine laterally flexed to contralateral side.

INTRODUCTION

A patient presenting with elbow pain can pose many challenges to the physiotherapist. In part, this may be due to the lack of clear aetiology in many circumstances. Also, due to the complexity of the human pain experience, a biopsychosocial approach to management is indicated. This chapter will explore this approach, using the multi-dimensional definition of pain given in Chapter 7.1, in response to the information provided in the case study.

ASSESSMENT FINDINGS

QUESTION 1

What are the components contributing to Miss NS's arm pain?

MULTIPLE COMPONENTS OF PAIN

As a starting point, it makes sense to review Miss NS's assessment, in order to identify the mechanisms underlying her pain. To support this discussion a search of the literature was performed, focusing on upper limb work related musculoskeletal disorders (WRMD), repetitive strain injury, non-specific arm pain, and lateral epicondylalgia (and variants in terminology: tennis elbow, and lateral epicondylitis (Waugh 2005 **C**)). There will be a comment about the relevance of labels such as 'acute' and 'chronic' pain, and about the use of the Yellow Flags approach to psychosocial assessment.

Threatening receptive input

Evidence of tissue damage would provide support for a nociceptive mechanism contributing to Miss NS's report of pain. However, from the assessment findings there is no convincing evidence of tissue damage. There has been some speculation about the repetitious action of keyboarding causing microtrauma and inflammation, but the existence of such microtrauma is not supported by the literature, as no inflammatory component has been identified (Awerbuch 2004 **C**; Davis 1999 **R**; Helliwell & Taylor 2004 **R**; Ireland 1998 **R**; Mitchell et al. 2000 **R**). While palpation findings in Miss NS indicated a focal area of exquisite tenderness, without other signs of an inflammation response it would be a broad assumption to conclude there was or had been a tissue injury. The fact that NSAIDs had no effect on symptoms reinforces this interpretation.

While unable to identify a nociceptive trigger related to tissue damage, there may still be a sensory component. As well as being tender, muscles were noted as having increased tone. Potentially this could cause pressure on surrounding tissues or on the muscle fibres themselves. If the pressure caused an excessive distortion of the tissues then the threshold required to trigger the mechanical nociceptors might have been reached. This is more likely to occur when the nervous system is in a sensitised state and when the activity of the muscles is at its greatest (such as in keyboarding or note taking). The increased blood flow increases the volume of the muscle, resulting in greater pressure on surrounding tissues. Indeed the pressure may create ischaemia (Helliwell & Taylor 2004 **R**) or a compartment-type syndrome (Pritchard et al. 2005 **A**). This increased muscle activity and resultant pressure might explain the loss of free movement of nerves, as noted in patients with non-specific arm pain (Greening et al. 2005 **A**). This can lead to neuropathic sensitivity in response to deformation or compression of neural tissue. Miss NS's response to neurodynamic testing might be indicative of this.

Muscle fatigue and delayed onset muscle soreness (DOMS) might also be considered as nociceptive inputs for Miss NS's perception of pain, but the mechanisms, at least for experimentally induced DOMS, appear to be distinguishable from pain in patients with lateral epicondylalgia (Slater et al. 2005 **A**).

State and structure of the nervous system

When considering potential influences on the sensitivity of the central nervous system it is necessary to include activity-dependent neuroplasticity. Repetitious or persistent neural activity patterns are likely to lead to cortical reorganisation, including enlarged or blurred representations (both motor and sensory) (Flor 2003 **R**; Robertson et al. 2003 **C**), which can lead to problems with motor control, and possibly to pain in stressful situations (for example, under excessive workload demands). Repetitious stimulation was found to exacerbate and prolong responses to noxious stimuli in a study comparing patients with arm pain and healthy controls (Montoya et al. 2005 **A**). The evidence from this study suggests that the enhanced responses are mainly evident in sensitised nervous systems, but there is some indication that repetition may lead to reduced thresholds in normal limbs. This is supported by research showing that continuing with repetitive work of more than 25 hours per week is a factor in poor prognosis (Waugh et al. 2004 **A**).

Finally, the reduced spontaneous activity demonstrated by Miss NS may be important. The sensory, proprioceptive, and visual inputs associated with lack of movement, and even the wearing of the elasticised tubular bandage, may be considered to be threatening receptive input, if the brain interprets them as signs of danger, damage or vulnerability. This depends to a large degree on Miss NS's prior experiences and learning. The state and structure of the nervous system will be influenced by these previous experiences and modified by both threatening receptive stimuli and internal beliefs and emotions. Some of this will be explored further in 7.3.

Internal beliefs and emotions

There is a close interdependence between beliefs and other cognitions and emotions, so it is not sensible to discuss them separately. High perceived stress levels, low mood, distress and anxiety, unhelpful thoughts about the cause of pain, a passive coping style (including catastrophising) and fear avoidance have all been identified as important risk factors for the development of a chronic pain problem (Overmeer et al. 2004 **A**). Research into work-related upper limb pain also suggests that many of these factors may be involved in the onset of pain (Awerbuch 2004 **R**; Helliwell & Taylor 2004 **R**; Spence & Kennedy 1989 **A**). While cognitive and emotional issues are not commonly considered until a problem becomes chronic, the number of factors contributing in the acute stage can lead to a complexity that demands a multi-dimensional approach from the outset.

The usefulness of the terms 'acute' and 'chronic' pain must therefore be questioned. The assumption is that the longer someone has pain, the more disability he/she will have. However, this is untrue. There are patients living with chronic pain demonstrating low use of health resources (Elliott et al. 1999 **A**), working effectively despite pain (Blyth et al. 2003 **A**), and with low levels of disability (Blyth et al. 2003 **A**). Level of disability is not so much an issue of chronicity as one of complexity. A person who has had pain for a long time may be well adjusted to it and not be disabled by it at all; this is what pain management programmes strive for. Equally, people can present with a new pain (for example, simple indigestion perceived as cardiac pain) and be very disabled. Of course, where an individual does not adapt well to an ongoing pain condition, disability will reflect cognitive and behavioural responses to the pain (as outlined above), which, if unchecked, can be expected to become more complex with time. So not only can these factors influence the perception of pain, but when seen as maladaptive responses to persistent pain, they can be the main influence on level of disability.

The literature does attempt to define chronic pain. One author review suggests presentations of lateral epicondylalgia that last longer than four to six weeks should be described as chronic (Vincenzino et al. 2002 **C**). However, this does not correspond with the slightly ambiguous IASP definition of chronic pain cited recently in Van Leeuwen et al. (2006 **A**), which states chronic pain is '... pain experienced every day for three months over a six month period' (p. 161). Further, a study examining the influence of symptom duration on prognosis, suggested that three years was a more distinguishing time frame (Dunn & Croft 2006 **A**). This lack of clarity and the increasing support for psychological interventions in the early stages of pain (Linton 2005 **R**; Pincus et al. 2002 **A**; Sullivan & Stanish 2003 **A**) imply that health professionals should always consider all the contributing components and mechanisms of pain (such as its complexity) from the outset, regardless of chronicity.

Therefore, in order to assess Miss NS's pain it is important to assess the psychological and social influences from the first contact. That is, a biopsychosocial assessment is essential. Miss NS is seeking physiotherapy at a private practice without ready access to a multi-disciplinary team, but this does not preclude the therapist from providing a biopsychosocial assessment. Indeed there is some appropriate information provided in the assessment summary.

The importance of this information might be made clearer by mapping it against the categories described in the Yellow Flags approach (Kendall et al. 1997 **C**). This approach was developed for the psychosocial assessment of patients with acute low back pain and aims to identify risk of long-term disability and, in turn, behavioural treatment targets that might prevent long-term problems (Watson & Kendall 2000 **C**). It focuses on the impact of pain rather than on the cognitive-emotional component of the perception of pain, but it would be surprising if there was no overlap. In order to identify these factors, a standard questionnaire could be applied, or the assessment interview could focus on the key factors, including attitudes and beliefs about pain, behaviours, compensation issues, diagnostic and treatment issues, emotions, family, and work (Watson & Kendall 2000 **C**).

If the Yellow Flags approach (see Table 7.2.1) is used to interpret the information Miss NS has provided – and it seems to be increasingly applied to all patients with painful conditions, not just those with low back pain (Bope et al. 2004 **C**; Brox 2003 **C**; Turner & Dworkin 2004 **R**) – then some clear treatment targets arise. Certainly, she appears to have some unhelpful beliefs about tissue damage and views work-related activities as injurious. This view is extending to activities of personal care and threatens her independence. It is possible that these beliefs lead her to be more vigilant of her arm posture and movement, heightening the sensitivity of the nervous system through attentional focus. With regard to behaviour, she has remained working and living independently and persists in tasks despite pain (including driving to parents' house). However, her quality of movement and the wearing of the elastic bandage need to be addressed. There are no compensation issues but work clearly has its

Table 7.2.1. Summary of psychosocial factors predictive of poor outcome (yellow flags), with examples from the current case

Psychosocial Factors	Examples
Attitudes and Beliefs	Belief that pain is harmful.
Behaviours	Excessive reliance on use of elastic bandage.
Compensation	Not evident.
Diagnosis and Treatment Issues	Multiple diagnoses.
Emotions	Pain-related fear.
Family	Potential for reduced social support (for example, lives alone).
Work	Management unsupportive in current work environment.

problems, as she is feeling stressed and unsupported to the point of looking for an alternate job. From the information provided it is not clear if issues relating to family, such as their role in reinforcing attitudes and beliefs or behaviour, are significant. It is also unclear if Miss NS has any emotional contributors, although anxiety could be inferred and it would not be surprising if her mood was low. Finally, her referral to physiotherapy involved three possible diagnoses or labels. This potentially causes confusion, especially when a non-tennis playing patient is told they have 'tennis elbow'. The other two labels of tendinitis and repetitive strain injury are unlikely and misleading, respectively.

Before these labels are addressed, a note of warning: anecdotal reports from the clinical environment suggest that the Yellow Flags approach is being applied unhelpfully. Rather than being used as a meaningful part of assessment that is helpful in identifying treatment targets and guiding treatment selection, it is being used as a label itself (for example, the patient is 'full of yellow flags', or worse, 'a Yellow Flagger'). In the past, 'supratentorial' and 'psychosomatic' have likewise been used to identify patients with presentations that do not neatly fit into a tissue-based model of care. Those guilty of this would do well to read Main and Waddell's (1998 C) guiding comments about the misuse of Waddell's signs of maladaptive pain behaviour.

DIAGNOSIS

QUESTION 2

What is an appropriate label for Miss NS's arm pain?

Miss NS has been presented with three diagnoses or labels for her condition: 'tennis elbow', 'tendinitis', and 'repetitive strain injury'. This section will explore the latter two and presumes 'tennis elbow' is unhelpful to both health professionals and patients alike.

TENDINITIS VERSUS TENDINOPATHY

The image of an inflamed tendon after excessive repetitive movement is a seductive one. It is easy to conceive a structure moving repetitively reaching some limit where the structure will begin to breakdown. An assumption of overuse may follow. However, there is a risk when we create a model of what is going on, that we substitute the actual structures with familiar non-organic structures, or make assumptions about the nature of the tissues and processes involved. For example, the concept of wear and tear does not fit the structures internal to the body. Despite common assumptions, our joints do not wear out like a shoe. Research into the pathogenesis of joint degeneration points to a history of injury and an inadequacy of active repair processes, rather than a simple attribution to workload. 'Wear and inadequate repair' might be a more appropriate description, although patients may still be discouraged from performing beneficial weight-bearing exercise for their degenerative arthritis (McCarthy et al. 2004 **R**).

The evidence for tendon damage in common tendon pain supports the notion that processes other than tissue injury are involved. The research literature outlines an interesting search for the mechanism of pain in tendinopathy, and inflammation appears to be ruled out (Khan et al. 1999 **R**). Therefore, clinicians are advised strongly to avoid referring to tendon pain as tendinitis unless they have confirming histological evidence. Recent findings of abnormal vascularisation and malalignment of fibres (Khan et al. 1999 **R**) and overload of tensile tissues (Hamilton & Purdam 2005 **C**) are the current favoured hypotheses, although the nociceptive mechanisms (the sensory component of the pain neuromatrix) remain undetermined or unproven. Sensitivity of the nervous system seems to have been neglected in these discussions of tendon pain, as has the role of the cognitive and emotional dimensions of pain. Interestingly however, the most effective treatment is the use of high load eccentric contractions, resulting in reduced pain and return to function (Alfredson et al. 1998 **A**; Cook et al. 2000 **C**). Similar treatment has been promoted in the exquisitely painful Complex Regional Pain Syndrome Type 1 (Watson & Carlson 1987 **A**). An interpretation of these surprising outcomes is that by promoting an unguarded forceful movement, the clinician sends a message to the patient that their body is not vulnerable. Further, the inputs and outputs of the nervous system are normalised, which encourages less vigilance of somatosensory and nociceptor information. Maybe this treatment approach demonstrates neuroplastic desensitisation (or learning), rather than a tissue healing process. The role of neuroplasticity and sensitisation of the nervous system may be a key feature in the report of tendon pain.

REPETITIVE STRAIN INJURY TO NON SPECIFIC ARM PAIN

The second label to consider is 'repetitive strain injury'. According to Helliwell and Taylor (2004 **R**), the common sufferer of repetitive strain injury is 'a female office or production line worker, conscientious in her job, who develops forearm pain after a change in work practice, additional demands, or pressure from supervisors' (p. 438). They also describe a diffuse arm pain that can spread to shoulder and neck regions, with work tasks the main factor in exacerbation. Miss NS fits these descriptors well. However, the creation of the term 'repetitive strain injury' has been attributed to a trade union spokesperson (Awerbuch 2004 **R**) and would appear to be an inaccurate description of the pathological processes involved; not that they are well understood (Awerbuch 2004 **R**; Davis 1999 **R**; Helliwell & Taylor 2004 **R**; Macfarlane et al. 2000 **A**). Indeed, the inappropriateness of this label is highlighted by the action of the Royal Australasian College of Physicians, discouraging its use since 1986 (Helliwell & Taylor 2004 **R**).

Suggestions of new labels for this condition include 'non-specific diffuse forearm pain' (Helliwell & Taylor 2004 **R**) and the more general 'non-specific arm pain' (Greening et al. 2005 **A**), which are in line with the diagnosis by exclusion of 'lumbar spine pain of no known origin' (Merskey & Bogduk 1994 **C**), commonly described as 'non-specific low back pain' (NSLBP). As with NSLBP, the 'non-specific arm pain' label may not be that helpful for patients, but recognises the inadequacy of a tissue-based paradigm in painful conditions (Gifford 1998 **C**).

To assist with Miss NS's management, a label or working diagnosis that excludes an inflammatory process or specific structure (such as a tendon), and focuses instead on the perception of pain, would be appropriate. 'Non-specific diffuse forearm pain' is limited by its anatomical location, which does not match with Miss NS's description of her pain. Therefore, the preferred diagnosis would be 'non-specific arm pain'. This is not an uncommon label to select, as was demonstrated in the development of epidemiological criteria for upper limb soft-tissue disorders (Helliwell et al. 2003 **A**). Using consecutive new cases and evaluation criteria consisting of 30 variables, the findings demonstrate that non-specific upper limb disorder was more than twice as prevalent as any tissue-specific diagnostic group (for example, inflammatory arthritis; lateral epicondylitis; shoulder tendinitis).

While the 'non-specific arm pain' label might be the health professional's preference, there is one more factor that needs consideration and that is the benefit, or otherwise, of giving a patient a new label for their condition. Kouyanou et al. (1998 **A**) warn that explanations that do not indicate a source of pain can lead the patient to believe their pain is imaginary. Persisting with the label 'repetitive strain injury' may be more meaningful (if misleading) and at least will allow for potentially informative personal research into the condition. As stated previously, Miss NS's presentation fits the definition, even if the term does not match the pathogenesis. Whatever term is chosen, education about the condition is essential and should be the focus of the initial intervention.

TREATMENT

QUESTION 3

What is the best treatment for non-specific arm pain?

In response to Miss NS's biopsychosocial assessment, a brief problem list might be constructed as in Table 7.2.2. Please note that this representation does not allow

Table 7.2.2. Identified key treatment targets for physiotherapy from biopsychosocial assessment

Threatening Receptive Stimuli	Increased muscle tone and guarding posture (including right upper limb and cervicoscapular muscles). Sensory, proprioceptive and visual input interpreted as damaged or vulnerable limb.
Internal Beliefs and Emotions	Concerns and distress about tissue injury and prognosis. Workplace stress and anxiety.
State and Structure of Nervous System	Sensitised due to above factors. Abnormal afferent and efferent activity due to reduced movement.

for the interaction of factors or the potential impact of treatments on all aspects of the individual.

'HANDS ON' VERSUS 'HANDS OFF'

Influenced by the uncertain dichotomy of 'acute' and 'chronic' pain is the equally worrisome 'hands on' and 'hands off' with regards to treatment. Klaber Moffett and Mannion (2005 **R**) raise this as a treatment quandary for physiotherapists when managing patients with low back pain. However, it is doubtful that this dualism will promote the effective management of patients with multi-dimensional problems (Spence & Kennedy 1989 **A**). Creating treatment targets in response to a biopsychosocial assessment is a strong basis for dealing with the range of individual presentations likely to occur. It should also ensure a patient-centred approach. A decision made on the simple reasoning that someone has either an acute or chronic pain is likely in many cases to be misguided and ineffective.

EVIDENCE FOR TREATMENT

It is recommended that treatments are evidence-based. According to Sackett et al. (2000 **C**), an evidence-based approach comprises best research evidence, clinical experience and patient expectation.

The research evidence to support physical interventions in presentations similar to Miss NS's is scant. A recent systematic review of physical interventions for lateral elbow pain reported a lack of evidence for long-term effectiveness (Bisset et al. 2005 **R**), although several investigators conclude there is some support for the inclusion of manual therapy on the cervical spine (Paungmali et al. 2004 **A**; Vincenzino 2003 **R**). Cochrane reviews searching for evidence to support the use of deep transverse friction massage in 'tendonitis' (Brosseau et al. 2002 **R**) or use of orthotic devices in 'tennis elbow' (Struijs et al. 2002 **R**) concluded there was no definite support for either. Also, a Cochrane review of biopsychosocial management for upper limb pain identified just two appropriate studies (Karjalainen et al. 2000 **R**). Notably, the criteria for the review excluded the possibility that such management could be undertaken by a solo practitioner (Karjalainen et al. 2000 **R**). Evidence does support the use of cognitive behavioural therapy (CBT) for the management of chronic pain conditions (Klaber Moffett & Mannion 2005 **R**; Spence 1989 **A**; Spence & Kennedy 1989 **A**; Sullivan & Stanish 2003 **A**), but the majority is from research on low back pain. Finally, physical exercise has been shown to be of some benefit to people with fibromyalgia (Busch et al. 2002 **R**; Da Costa et al. 2005 **A**), which, according to Helliwell and Taylor (2004 **R**), is similar in nature to the non-specific arm pain as reported by Miss NS. Certainly, the evidence from the literature on tendinopathy suggests it may be worth exploring whether there is a role for eccentric loaded exercise.

Physical therapy

There is research that suggests manual therapy is a popular choice of treatment (Greenfield & Webster 2002 **A**). The conclusions of this survey, investigating physiotherapist treatment selection for chronic lateral epicondylitis, state a large number of physiotherapists (approximately 40 % of sample) used manipulation, of the elbow, only when other treatments had failed. Manipulation of the cervical spine has some support in the literature (Cleland et al. 2004 **A**) but its use risks reinforcing a passive coping approach, as well as potentially re-focusing Miss NS's health anxiety. The most popular treatments were progressive stretching, progressive strengthening, and deep transverse friction (Greenfield & Webster 2002 **A**). Regarding Miss NS's reduced movement and activity and increased muscle tone, these strategies may be beneficial in promoting relaxation and increased blood flow. Given the need to incorporate best evidence and the emphasis on actively involving the patient in cognitive-behavioural interventions, it would seem best to incorporate the stretching and strengthening into a home exercise programme and avoid deep transverse friction, which lacks research support.

This is not to say that performing assisted stretches or applying massage would always be detrimental. With the right emphasis, such a session might be educational for the patient in terms of the vigour with which techniques can be safely applied, the demonstration of appropriate end-feel, and if done well, the promotion of the physiotherapist as a movement facilitator, rather than a healer. There should also be some beneficial tissue effects, including normalising of the experience of the nervous system.

Cognitive-behavioural interventions

Education

Moseley (2003b **A**) used an educational intervention on chronic low back pain patients and demonstrated that simple physical outcome measures can be changed in response to cognitive changes. A key feature of this was improvement in catastrophising score. No formal measure of catastrophising was reported in Miss NS's assessment, but it is possible that she believes that using her painful arm will result in a need for surgery, as happened to one of her colleagues. Effective education would address these concerns and is arguably the best evidence-based intervention for non-specific upper limb pain. Physiotherapists are well placed to provide such education, which, delivered in conjunction with exercise, can be used to directly challenge the patient's beliefs about activity and damage.

Active versus passive treatment

Miss NS's treatment plan needs to be further modified by the clinician's expertise and her own expectations. The former is obviously difficult to describe meaningfully here as each physiotherapist has an individual experience, knowledge and skills set. However, individualising the treatment plan will be the focus of the remainder of this chapter.

As the primary goal of treatment of non-specific pain is to promote and encourage return to normal activity (Harding & Watson 2000 **C**; Harding & Williams 1998 **C**; Klaber Moffett & Mannion 2005 **R**), it is essential to select treatments that support this. So-called 'hands on' treatments might still be applied in a manner that promotes patient activity. However, they would need to be adjuncts to more active strategies and be supported by education of the patient. In providing treatment that is delivered entirely by the therapist (such as manual therapy), there is a risk of promoting a dependent relationship. Such a treatment approach takes responsibility away from the patient and can lower self-efficacy for self-management and promote the unhelpful attitude of waiting for the pain to be taken away (Nicholas & Sharp 1999 **C**).

In contrast, by giving Miss NS an active role in her treatment, there will be positive effects both physically and psychologically. A first step will be to clearly report the assessment findings so that a meaningful discussion can be pursued. This should begin with a statement confirming the legitimacy of Miss NS's pain report. Next inform her about the results of the physical examination, which did not identify any reliable sign of tissue injury but did identify movement anomalies associated with guarding behaviour. This establishes a rationale for movement rather than immobility.

Facilitating problem solving

Through discussion of these findings the physiotherapist can help address unhelpful beliefs. Education about the research into non-specific arm pain, which concludes that symptoms are not directly related to intensity or quantity of repetition, will be helpful in adapting Miss NS's beliefs about work. Also, informing her that there is no evidence of tissue damage in many similar cases will encourage her to question her belief about limiting activity and her fear of a worsening prognosis. Highlighting the strongly implicated role of workplace stress is also essential and will prompt Miss NS to assess and address this.

The discussion should also aim to raise awareness of the plastic nature of the nervous system and how it can respond to inactivity and guarding by cortical reorganisation and increased sensitivity (Flor 2003 **R**; Robertson et al. 2003 **R**). This level of discussion about the neurophysiology of pain is supported in the low back pain literature (Moseley 2003b **A**). In order for the discussion to be most effective, the physiotherapist needs to avoid simply giving information and instead assist Miss NS in raising her own questions and forming her own conclusions.

Using behaviour to challenge beliefs

This may not all happen in one session, and indeed it would be unrealistic to think that such a discussion will be powerful enough to change beliefs entirely. Therefore, the physiotherapist should encourage Miss NS to test her own hypotheses using behavioural experiments. For example, by clarifying the role of an elastic tubular bandage (the control of swelling) and pointing out its potential negative influence on movement and hypervigilance, the therapist will encourage Miss NS to question its value when no swelling is present. In response to this, she may decide on a strategy to reduce the wearing of the bandage herself. If the strategy is successful, the realisation that her original belief was unhelpful will be reinforced. If the strategy she uses is unsuccessful – or if she cannot think of a strategy – then collaborative goal setting will enable a realistic plan for her to reduce and terminate its use (Arnetz et al. 2004 **A**). Essentially, while education and discussion may provide the prompt to try and change unhelpful beliefs, successfully performing a behaviour that is incompatible with the unhelpful belief will actually cause the belief to change (Prochaska et al. 1992 **R**).

Summary of treatment

An appropriate treatment approach for Miss NS would consist of information sharing and discussion, independent and/or collaborative goal setting, strategies to promote normal upper limb and cervical posture and movement, normalising nervous system experiences, and addressing workplace stress. Treatment should focus on actively involving Miss NS, promoting self-management and providing opportunities for her to use her own problem solving skills, especially in the workplace, where she is likely to be intimately aware of potential stressors. Ideally, the aim of treatment will be to integrate physical outcomes relating to movement with psychological outcomes (improved self-efficacy and coping), and with socio-environmental outcomes (reduction in workplace stressors). Thus, threatening receptive inputs, unhelpful beliefs and negative emotions can all be modified, and concurrently the state of the nervous system will be normalised.

This intervention reflects conclusions by Stephenson (2002 **C**). He proposes a new paradigm for physiotherapy, where psychological and social factors are not seen as 'confounding variables', but are rated and addressed equally with more traditional physical targets (p. 254). In line with this, follow-up sessions should include reviewing goals, reviewing beliefs and attitudes, and implementing strategies to match change in physical performance (Moseley 2003a **C**).

CONCLUDING REMARKS

Bishop and Foster (2005 **A**) suggest physiotherapists may be under-confident or have limited knowledge and skills to apply a biopsychosocial approach. It has also been found that physiotherapists need to develop collaborative goal-setting skills (Gladwell 2006 **C**; Parry 2004 **A**). It is important that physiotherapists develop these areas in order to effectively employ a patient-centred approach to care. This would seem most important when patients present with pain that does not have a clear nociceptive component. Consideration of other threatening receptive input is required, along with attention to unhelpful beliefs and negative emotions. Further, neglecting these factors in any individual, even in the early stages of an injury, may detrimentally affect the eventual outcome (Stephenson 2002 **C**). Where the physiotherapist identifies problems beyond his or her scope of skills and knowledge, a clinical psychologist should be involved – earlier, not later, if possible.

In addition, physiotherapists must work to understand their own beliefs about pain and nervous system sensitivity, and the influence of these beliefs on their communication (Daykin 2006 **C**). This reflective approach will reduce the risk of misinterpretation of a patient's pain report and promote therapeutic alliance. It is also important to highlight the fact that reflection, rather than training, might alert physiotherapists to a range of unrecognised skills they have in this area. Most physiotherapists already, perhaps unconsciously, employ strategies that directly or indirectly influence cognitions and emotions in order to engage, motivate and educate patients. Arguably, there is a case for re-labelling, re-interpreting and refining these strategies, rather than having to learn something new.

People can problem solve, thought-challenge and adopt new behaviours independently, without professional guidance (Prochaska et al. 1992 **R**). Sometimes patients may simply need appropriate information and a supported opportunity to initiate this. As physiotherapists, we need to ensure we are promoting and not inhibiting this independent patient-centred approach. It would seem that many patients with non-specific pain conditions benefit from interventions that promote independence. Approaches and interactions which depend on the therapist may interfere with this and in some cases be considered iatrogenic.

NOTE

While the following term and definition may seem facetious, it highlights a concern that when a clinician is faced with complexity beyond their knowledge and skills, the perceived obligation to do something may prevail:

Threatment – the menacing behaviour of a health professional who feels compelled to do something to a patient even though it is not in the patient's best interest. (Roland & Jones, personal communication).

ACKNOWLEDGEMENTS

With thanks to Helen Skehan at Physiosolutions, Heidi Roland, Ingrid Wilson and Miss NS.

REFERENCES

Alfredson H, Pietila T, Jonsson P, Lorentzon R (1998) Heavy-load eccentric calf muscle training for the treatment of chronic Achilles tendinosis. *American Journal of Sports Medicine* **26**: 360–366.

Arnetz JE, Bergstrom AK, Franzen Y, Nilsson H (2004) Active patient involvement in the establishment of physical therapy goals: effects on treatment outcome and quality of care. *Advances in Physiotherapy* **6**(2): 50–69.

Awerbuch M (2004) Repetitive strain injuries: has the Australian epidemic burnt out? *Internal Medicine Journal* **34**: 416–419.

Bishop A, Foster NE (2005). Do physical therapists in the United Kingdom recognise psychosocial factors in patients with acute low back pain? *Spine* **30**(11): 1316–1322.

Bisset L, Paungmali A, Vicenzino B, Beller E (2005) A systematic review and meta-analysis of clinical trials on physical interventions for lateral epicondylalgia. *British Journal of Sports Medicine* **39**(7): 411–422.

Blyth FM, March LM, Cousins MJ (2003) Chronic pain-related disability and use of analgesia and health services in a Sydney community. *Medical Journal of Australia* **179**(2): 84–87.

Blyth FM, March LM, Nicholas MK, Cousins MJ (2003) Chronic pain, work performance and litigation. *Pain* **103**(1–2): 41–47.

Bope ET, Douglass AB, Gibovsky A, Jones T, Nasir L, Palmer T et al. (2004) Pain management by the family physician: the family practice pain education project. *Journal of the American Board of Family Practice* **17**: S1–12.

Brosseau L, Casimiro L, Milne S, Robinson VA, Shea BJ, Tugwell P et al. (2002) Deep transverse friction massage for treating tendinitis. *Cochrane Library* **4** http://www.thecochranelibrary.com CD003528.

Brox JI (2003) Regional musculoskeletal conditions: shoulder pain. *Best Practice and Research in Clinical Rheumatology* **17**(1): 33–56.

Busch A, Schachter CL, Peloso PM, Bombardier C (2002) Exercise for treating fibromyalgia syndrome. *Cochrane Library* **2** http://www.thecochranelibrary.com CD003786.

Cleland JA, Whitman JM, Fritz JM (2004) Effectiveness for manual physical therapy to the cervical spine in the management of lateral epicondylalgia: a retrospective analysis. *Journal of Orthopaedic and Sports Physical Therapy* **34**(11): 713–724.

Cook J, Khan K, Maffuli N, Purdham C (2000) Overuse tendinosis, not tendinitis: applying the new approach to patella tendinopathy. *Physician and Sports Medicine* **28**(6): 31–46.

Da Costa D, Abrahamowicz M, Lowenstyn I, Bernatsky S, Drista M, Fitzcharles M-A et al. (2005) A randomized clinical trial of an individualized home-based exercise programme for women with fybromyalgia. *Rheumatology* **44**: 1422–1427.

Davis TR (1999) Do repetitive tasks give rise to musculoskeletal disorders? *Occupational Medicine* **49**(4): 257–258.

Daykin A (2006) Communication and assessment: message received and understood. In: Gifford L (ed.) *Topical Issues in Pain 5* Falmouth: CNS Press.

Dunne KM, Croft PR (2006) The importance of symptom duration in determining prognosis. *Pain* **121**: 126–132.

Elliott AM, Smith BH, Penny KI, Smith WC, Chamber WA (1999) The epidemiology of chronic pain in the community. *Lancet* **354**: 1248–1252.

Flor H (2003) Cortical reorganisation and chronic pain: implications for rehabilitation. *Journal of Rehabilitation Medicine* **41** Suppl.: 66S–72S.

Gifford LS (1998) Pain, the tissues and the nervous system: a conceptual model. *Physiotherapy* **84**(1): 27–36.

Gladwell P (2006) A practical guide to goal-setting. In: Gifford L (ed.) *Topical Issues in Pain 5* Falmouth: CNS Press.

Greenfield C, Webster V (2002) Chronic lateral epicondylitis. *Physiotherapy* **88**(10): 578–594.

Greening J, Dilley A, Lynn B (2005) In vivo study of nerve movement and mechanosensitivity of the median nerve in whiplash and non-specific arm pain patients. *Pain* **115**: 248–253.

Hamilton B, Purdam C (2004) Patellar tendinosis as an adaptive process: a new hypothesis. *British Journal of Sports Medicine* **38**: 758–761

Harding V, Watson P (2000) Increasing activity and improving function in chronic pain management. *Physiotherapy* **86**(12): 619–629.

Harding V, Williams AC de C (1998) Activities training: integrating behavioral and cognitive methods with physiotherapy in pain management. *Journal of Occupational Rehabilitation* **8**(1): 47–60.

Helliwell PS, Taylor WJ (2004) Repetitive strain injury. *Postgraduate Medicine Journal* **80**: 438–443.

Helliwell PS, Bennett RM, Littlejohn G, Muirden KD, Wigley RD (2003) Towards epidemiological criteria for soft-tissue disorders of the arm. *Occupational Medicine* **53**(5): 313–319.

Ireland DC (1998) Australian repetition strain phenomenon. *Clinical Orthopaedics and Related Research* **351**: 63–73.

Karjalainen K, Malmivaara A, van Tulder M, Roine R, Jauhiainen M, Hurri H et al. (2000) Biopsychosocial rehabilitation for upper limb repetitive strain injuries in working age adults. *Cochrane Library* **3** http://www.thecochranelibrary.com CD002269.

Kendall NAS, Linton SJ, Main CJ (1997) *Guide to Assessing Psychosocial Yellow Flags in Acute Low Back Pain: Risk Factors in Long-term Disability and Work Loss* Wellington, New Zealand: Accident Rehabilitation and Compensation Insurance Corporation of New Zealand, and the National Health Committee, Ministry of Health.

Khan KM, Cook JL, Bonar F, Harcourt P, Astrom M (1999) Histopathology of common tendinopathies. Update and implications for clinical management. *Sports Medicine* **27**(6): 393–408.

Klaber Moffett J, Mannion AF (2005) What is the value of physical therapies for back pain? *Best Practice and Research* **19**(4): 623–638.

Kouyanou K, Pither CE, Rabe-Hasketh S, Wessely S (1998) A comparative study of iatrogenesis, medication abuse, and psychiatric morbidity in chronic pain patients with and without medically explained symptoms. *Pain* **76**: 417–426.

Linton SJ (2005) Do psychological factors increase the risk for back pain in the general population in both a cross-sectional and prospective analysis? *European Journal of Pain* **9**(4): 354–361.

Macfarlane GJ, Hunt IM, Silman AJ (2000) Role of mechanical and psychosocial factors in the onset of forearm pain: prospective population based study. *British Medical Journal* **321**: 1–5.

Main CJ, Waddell G (1998) Behavioural responses to examination: a reappraisal of the interpretation of 'non-organic signs'. *Spine* **23**(21): 2367–2371.

McCarthy CJ, Mills PM, Pullen R, Richardson G, Hawkins N, Roberts CR et al. (2004) Supplementation of a home-based exercise programme with a class-based programme for people with osteoarthritis knees: a randomised controlled trial and health economic analysis. *Health Technology Assessment* **8**, no. 46.

Merskey H, Bogduk N (1994) *Classification of Chronic Pain: descriptions of chronic pain syndromes and definition of chronic pain terms* (2 edn) Seattle: IASP Press.

Mitchell S, Cooper C, Martyn C, Coggon D (2000) Sensory neural processing in work-related upper limb disorders. *Occupational Medicine* **50**(1): 30–32.

Montoya P, Pauli P, Batra A, Wiedemann G (2005) Altered processing of pain-related information in patients with fibromyalgia. *European Journal of Pain* **9**(3): 293–303.

Moseley GL (2003a) A pain neuromatrix approach to patients with chronic pain. *Manual Therapy* **8**(3): 130–140.

Moseley GL (2003b) Unravelling the barriers to reconceptualisation of the problem of chronic pain: the actual and perceived ability of patients and health professionals to understand neurophysiology. *Journal of Pain* **4**: 184–189.

Nicholas MK, Sharp TJ (1999) A collaborative approach to managing chronic pain. *Modern Medicine of Australia* October: 26–34.

Overmeer T, Linton SJ, Boersma K (2004) Do physical therapists recognise established risk factors? Swedish physical therapists' evaluation in comparison to guidelines. *Physiotherapy* **90**(1): 35–41.

Parry R (2004) Communication during goal-setting in physiotherapy treatment sessions. *Clinical Rehabilitation* **18**: 668–682.

Paungmali A, O'Leary S, Souvlis T, Vincenzino B (2004) Naloxone fails to anatagonise initial hypoalgesic effect of a manual therapy treatment for lateral epicondylalgia. *Journal of Manipulative and Physiological Therapeutics* **27**(3): 180–185.

Pincus T, Vlaeyen JWS, Kendall NAS, Von Korff NR, Kalauokalani DA, Reis S (2002) Cognitive-behavioral therapy and psychosocial factors in low back pain: directions for the future. *Spine* **27**(5): 133E–138E.

Pritchard MH, Williams RL, Heath JP (2005) Chronic compartment syndrome, an important cause of work-related upper limb disorder. *Rheumatology* **44**: 1442–1446.

Prochaska JO, DiClemente CC, Norcross JC (1992) In search of how people change: applications to addictive behaviours. *American Psychologist* **47**(9): 1102–1114.

Robertson EM, Theoret H, Pascual-Leone A (2003) Skill learning. In: Boniface S, Ziemann U (eds) *Plasticity in the Human Nervous System* Cambridge: Cambridge University Press, pp. 107–134.

Sackett DL, Strauss SE, Richardson WS, Rosenberg W, Haynes RB (2000) *Evidence-Based Medicine: how to practice and teach EBM* (2 edn) London: Churchill-Livingstone.

Slater H, Arendt-Nielsen L, Wright A, Graven-Nielsen T (2005) Sensory and motor effects of experimental muscle pain in patients with lateral epicondylalgia and controls with delayed onset muscle soreness. *Pain* **114**: 118–130.

Spence SH (1989) Cognitive-behaviour therapy in the management of chronic occupational pain of the upper limbs. *Behaviour Research and Therapy* **27**(4): 435–446.

Spence SH, Kennedy E (1989) The effectiveness of a cognitive-behavioural treatment approach to work-related upper limb pain. *Behaviour Change* **6**(1): 12–23.

Stephenson R (2002) The complexity of human behaviour: a new paradigm for physiotherapy. *Physical Therapy Reviews* **7**: 243–258.

Struijs PAPAA, Arola H, Assendelft WJJ, Buchbinder R, Smidt NN, van Dijk CN (2002) Orthotic devices for the treatment of tennis elbow. *Cochrane Library* **1** http://www.thecochranelibrary.com CD00182.

Sullivan MJL, Stanish WD (2003) Psychologically based occupational rehabilitation: the pain-disability prevention program. *Clinical Journal of Pain* **19**(2): 97–104.

Turner JA, Dworkin SF (2004) Screening for psychosocial risk factors in patients with chronic orofacial pain: recent advances. *Journal of American Dental Association* **135**(8): 1119–1125.

Van Leeuwen MT, Blyth FM, March LM, Nicholas MK, Cousins MJ (2006) Chronic pain and reduced work effectiveness. *European Journal of Pain* **10**(2): 161–166.

Vincenzino B (2003) Lateral epicondylalgia: a musculoskeletal physiotherapy perspective. *Manual Therapy* **8**(2): 66–79.

Vincenzino B, Souvlis T, Wright A (2002) Musculoskeletal pain. In: Strong J, Unruh AM, Wright A, Baxter GD (eds) *Pain: a textbook for therapists* London: Harcourt.

Watson HK, Carlson L (1987) Treatment of reflex sympathetic dystrophy of the hand with an active 'stress loading' program. *Journal of Hand Surgery American Volume* **12**: 779–785.

Watson P, Kendall N (2000) Assessing psychosocial yellow flags. In: Gifford L (ed.) *Topical Issues in Pain 2* Falmouth: CNS Press.

Waugh EJ (2005) Lateral epicondylalgia or epicondylitis: what's in a name? *Journal of Orthopaedic and Sports Physical Therapy* **35**(4): 200–202.

Waugh EJ, Jaglal SB, Davis AM (2004) Computer use associated with poor long-term prognosis of conservatively managed lateral epicondylalgia. *Journal of Orthopaedic and Sports Physical Therapy* **34**(12): 770–780.

7.3   Recurrent Lumbar Pain after Failed Spinal Surgery

LESTER JONES AND AUDREY WANG

CASE REPORT

BACKGROUND

Mr CP is a 30 year old man living with his fiancée. He has recently started work as a trainee solicitor in a small law practice. His workplace activities include keyboarding, use of a computer-mouse, the carrying and filing of legal paperwork, and meeting clients. While it is a new job, he is settling in well and looking forward to his new career.

He has a two year history of back pain and has had both invasive and non-invasive treatments to try to resolve it: a partial lumbar discectomy, which he took some time to recover from; a nerve block; and manipulation of 'facet joints' by a physiotherapist, which he reported as most effective. Following this treatment he was relatively pain free. He gradually returned to his sporting activities including gym, social rugby and football, and reported being unrestricted during these activities. However, during one game of football he felt discomfort and some stiffness in his back. He played on but the next day he noticed a dramatic increase in back stiffness. Believing that he had damaged the same or an adjacent lumbar disc, he reduced all unnecessary activity, especially anything that involved bending and twisting. He attended a private physiotherapy clinic for assessment 12 weeks after this game.

Mr CP says that because of pain he has to push himself at work sometimes, as it is a busy practice. He also reports leaving work early on occasions, often when sitting becomes too uncomfortable, and he has even had to take some days off due to periods of increased pain. At the time of physiotherapy assessment, he had accumulated a total of 12 days off work, including five days off for an unrelated chest infection. His work has an official policy of reviewing employees' performances if they take more than 15 days off work. He is a little worried about it but states that his boss has been supportive of him up to this time.

Mr CP has an upcoming performance appraisal, as part of a career structure, and the firm requires evidence of active participation in billing clients. He is concerned that

Recent Advances in Physiotherapy.   Edited by C. Partridge
© 2007 John Wiley & Sons, Ltd

his reduced attendance will affect his capacity to do this, and also his job promotion prospects.

While work is largely unaffected, he has stopped all sporting activities and developed an increasingly dependent role in activities at home.

MEDICAL DIAGNOSIS

None available.

ASSESSMENT

Initial presentation to physiotherapy

- Reports symptoms identical to those felt prior to partial discectomy.
- New job includes health insurance. Plans to use this to pay for MRI to review disc integrity.
- Some time off work.
- Wants review/opinion by physiotherapist.
- Walking tolerance is unaffected but sitting tolerance is reduced.
- Movement involving bending and twisting is painful.
- Avoidant of all activities that will put his 'disc' at risk.
- No dysaesthesias or referred pain.

On examination

- No obvious restriction in gait or stand-to-sit-to-stand.
- Back and upper limb muscles well developed and no sign of wasting.
- Balance and co-ordination of limbs appears normal.
- Palpation.
- Increased muscle tone around lumbar region bilaterally.
- Diffuse tenderness reported upper to mid lumbar.
- Joint movement (quality and range of physiological):
  - Lumbar spine – reduced speed and guarding, with flexion and then deviation into left lateral flexion/rotation from 40 degrees; reduced speed and guarding into rotation to left and right; all movement greater than $3/4$ range.
- Joint movement (quality and range of accessory):
  - Lumbar spine – generally stiff, especially middle and lower region, and painful end of range (central and unilateral).
- Muscle extensibility:
  - Reduced in erector spinae.
  - Reduced in gluteals, right more so than left.
  - Reduced in hip flexors.
- Neurological tests for sensation and reflexes normal. Passive straight leg raise restricted: 30 degrees right, 65 degrees left.

INTRODUCTION

A biopsychosocial approach conceptualises the person's pain experience as having the potential to be influenced by a number of factors. These factors include attitudes and beliefs, amount of psychological distress, illness behaviour, and social environment. This approach is increasingly recommended for managing both acute and chronic low back pain and preventing the transition between the two (Airaksinen et al. 2005 **R**; Burton et al. 2004 **R**; Kendall et al. 1997 **C**; Klaber Moffett & Mannion 2005 **R**; van Tulder et al. 2004 **R**).

The way a patient perceives their physical injury potentially has as much influence as the injury itself in determining either a full recovery or subsequent development of chronic disability. This may be particularly pertinent when pain recurs or persists. A patient might say 'Oh, I have developed a new back problem,' when in fact this is the third episode of a pre-existing back problem. It will be important to take a step back and view the bigger picture. A recurrence of pain at the site of an old injury should not be treated in the same way as an acute injury. Concepts such as central sensitisation and cortical reorganisation should be incorporated into clinical reasoning, especially if pain or tenderness are the only signs of tissue damage that are present. Terms such as 'acute-on-chronic' perhaps reflect a reasoning error that pain can only occur in response to more tissue damage.

Performing a biopsychosocial assessment may uncover fears and unhelpful beliefs, or social difficulties that contribute to a heightened sensitivity of the nervous system. Obviously, when these fears, beliefs and difficulties have a serious impact on an individual, social work and psychology professionals should be involved. Increasingly however, in less serious cases, physiotherapists are attending to these non-physical factors with the aim of promoting self-management and reducing long-term disability. The information provided on Mr CP suggests that a physiotherapist could take such a route in his management.

ASSESSMENT FINDINGS

QUESTION 1

What are the components contributing to Mr CP's low back pain?

According to European guidelines, the priority for assessment is to exclude non-spinal pathology, serious spinal pathology and nerve root pain (van Tulder et al. 2004 **R**). It can reasonably be established with the limited information provided that Mr CP's back pain is of a non-serious, non-specific type.

MULTIPLE COMPONENTS OF PAIN

It is sensible to start with a review of Mr CP's assessment, in order to identify the mechanisms underlying his pain.

Research evidence supporting this discussion was obtained by searching the literature, with a focus on low back pain and failed back surgery syndrome, as well as fear avoidance behaviour in chronic pain conditions.

The important role of neuroplasticity in altering nervous system sensitivity will be introduced here. Consideration of psychosocial factors will expand beyond the yellow flags described in Chapter 7.2 to include occupational factors: blue flags and black flags. A review of issues associated with fear avoidance behaviour in patients with low back pain will be presented as well.

Complex not chronic pain

'Chronicity', as discussed in Chapter 7.2, may not be an informative label for either the patient's abilities or for treatment selection. Noting the complexity of the patient's presentation is much more valuable.

In Mr CP's case there are a number of factors that might lead us to consider his pain as complex. First, he is certain it is a recurrence of previous symptoms and he believes he has damaged the intervertebral disc. Second, he has negative expectations about outcome. The initial injury had a big impact on his life, requiring surgery and additional rehabilitation. He anticipates a similar prognosis with this recurrence and is very concerned about causing more damage. Already it is impacting on normal movement and his new job and home life. Third, the physiological component of pain is unlikely to be straightforward. Notably, it is 12 weeks since these symptoms recurred, which for an otherwise healthy person provides adequate time for healing and repair of most tissue damage, and resolution of inflammation processes. Therefore it is very likely that the predominant nociceptive influence is the lowering of thresholds via central nervous system sensitivity. A thorough physical examination, including examination for red flags, and a review of psychosocial risk factors for long-term back pain (yellow flags) are essential to managing this complex presentation effectively.

Threatening receptive input

Mr CP reports a previous history of intervertebral disc damage and surgery. Due to the time elapsed since the recurrence of back symptoms, it would seem reasonable to believe that any tissue damage and resultant inflammation would be resolved by now. Therefore simple, local mechanical and chemical nociception are unlikely to be key in his perception of pain. This presumes he does not have any concurrent pathology that might delay healing, such as diabetes. It also presumes that he has not re-injured tissue in the last 12 weeks. His description of limited activity and movement makes re-injury unlikely.

It is possible some disc material or loose body is interfering with or compressing joint structures or nerve tissue (Miller et al. 2005 C), but from the assessment information there is no need to be concerned about spinal or nerve root involvement, and no reason to jump to this conclusion.

It is necessary to look for other triggers. The modified proprioceptive and visual sensory input Mr CP's brain is receiving as a result of his guarded movement may be being interpreted as a sign that something is wrong or is damaged. In essence, such input may be being treated as threatening receptive input. If he perceives his back to be vulnerable then his nervous system could already be sensitised (see below), and his individual pain neuromatrix (Moseley 2003 **C**) more susceptible to activation by these normally non-noxious sensory inputs.

The state and structure of the nervous system

Neuroplasticity, implicit in learning, is an adaptive process responding to meaningful sensory input and behaviourally-relevant activity (Flor 2003 **R**). Repetition, attention, and the difficulty of a task have been reported as influences on the excitatory and structural changes involved in neuroplasticity (Classen & Cohen 2003 **C**). Central and peripheral neuroplasticity can promote nervous system sensitivity to pain. Excitatory changes occur in response to tissue damage, via chemically-mediated changes to nociceptive thresholds, by activating neurons that are dormant prior to injury, and by making changes in inhibition centrally (for example NMDA receptor). Persistent pain conditions appear to be associated with structural re-organisation in the cortex. This potentially can result in a pain 'memory trace', which can be activated in the absence of peripheral stimuli (Flor 2003 **R**, p. 67). Although the mechanisms for this are not clear (Moseley 2006 **C**), it is unlikely to be simply that the patient has had pain for a long time. For example, Robertson, Theoret and Pascual-Leone (2003 **C**) hypothesise that high sensory demand can lead to faulty processing of sensory-motor information, leading to this pain sensitivity. This reinforces the need to consider the complex causes of pain, including maladaptive learning experiences, rather than focusing on chronicity.

It is possible that Mr CP's nervous system is undergoing both excitatory and structural changes. Neuroplasticity is activity dependent, so Mr CP's relative inactivity may have led the nervous system to respond by modifying synapses or reducing the potency of corticomotor patterns for unused movement. In addition there is likely to be a contribution to nervous system sensitivity from Mr CP's beliefs and emotions. If he is very concerned about re-injury then the attention or vigilance to sensory information relating to his back will be heightened. Based on Flor's (2003 **R**) work, this can drive cortical reorganisation such that somatosensory representation of the back is enlarged, leading to increased reactivity to tactile, or other non-noxious but potentially threatening stimuli. (This may underpin the pain and behaviour seen in patients who show a high fear of pain. Passive or 'hands on' treatments that target the painful area may also reinforce a maladaptive neuroplasticity.)

Mr CP's nervous system might already have been sensitised from the insult of the initial injury and/or the partial discectomy, especially given no resolution of the pain was immediate and his activity was restricted for some prolonged time after the surgery. His lumbar region would have been his focus, and attention to that region was reinforced by the hands on treatment he was receiving. This focus may have heightened the response of his primary somatosensory and primary motor cortex to threatening receptive stimuli, via altered cortical representations.

A recent prospective investigation identified that a combination of physical (work postures and activities) and psychosocial (fear related to pain) factors best predicted those who developed disabling low back pain (van Nieuwenhuyse et al. 2006 **A**). The physical factors involved in work were measured by response to standard items, rather than observation of work practice. It might be argued that those who rated these items highly had an already heightened attention to particular work postures. A hypervigilant nervous system – resulting from heightened attention – may be predisposed to triggering brain activity that leads to a perception of pain (activation of pain neuromatrix: see Chapter 7.2).

In other words, the nervous systems of individuals with fears about pain and tissue injury may be more alert to potentially vulnerable postures and more sensitive in reacting to sensory stimuli. The result is a primed nervous system that is more likely to create the perception of pain.

Internal beliefs and emotions

Fear avoidance behaviour

A strong body of literature supports the role of pain-related fear in subsequent disability in patients with low back pain (de Jong et al. 2005 **A**; Peters et al. 2005 **A**; Storheim et al. 2005 **A**; Swinkels-Meewisse et al. 2006 **A**; van Nieuwenhuyse et al. 2006 **A**; Vlaeyen & Linton 2000 **R**; Vlaeyen et al. 2002 **A**; von Korff et al. 2005 **A**). There is some evidence that in acute or sub-acute presentations, pain-related fear may not be a valuable predictor (Sieben et al. 2005 **A**) or a valuable treatment target (Jellema et al. 2005 **A**). However, the authors of these studies propose methodological explanations for their failure to identify the importance of pain-related fear.

Fear-avoidance behaviour is well represented by a model developed by Vlaeyen and colleagues, based on Letham's model (Vlaeyen & Linton 2000 **R**) (see Figure 7.3.1). While self-efficacy for self-management of pain is not represented on the model, it could be expected to feature in both the 'vicious' cycle of fear-avoidance (low self-efficacy) and the path to recovery (high self-efficacy) (Ashgari & Nicholas 2001 **A**). Recently, low self-efficacy has been associated with a vulnerable personality-type that may be predisposed to passive coping styles (Ashgari & Nicholas 2006 **A**). This would include avoidance and catastrophising as per the model.

Fear of re-injury is a significant issue in Mr CP's presentation. His belief about the nature of his condition (disc lesion) is likely not only to guide his behaviour but also to lead to cognitive and emotional sequelae. As a result, assessment and treatment decisions can be derived from the aforementioned model (see Figures 7.3.2a, 7.3.2b). The coping strategies that Mr CP has adopted in response to the recurrent episode of pain need to be evaluated as either helpful confronting (active) strategies or unhelpful avoiding (passive) strategies. Coping style has been shown to be an important determinant of level of participation (Burton et al. 1995 **A**; Linton 2005 **A**;

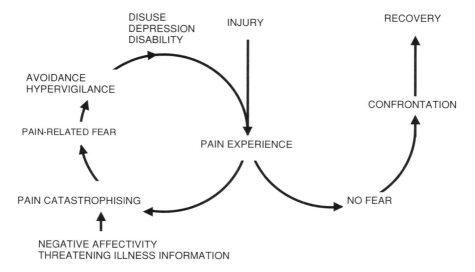

Figure 7.3.1. The 'fear' avoidance model (Vlaeyen & Linton **R** 2000, p. 329). Reproduced by kind permission of the International Association for the Study of Pain.

van Tulder et al. 2004 **R**). From the information provided, Mr CP appears to use a passive approach.

Catastrophising

Research places great emphasis on the role of catastrophising in the pain experience (Moseley 2004 **A**; Peters et al. 2005 **A**; Sullivan et al. 2004 **A**; Vlaeyen et al. 2002 **A**). Mr CP would appear to be catastrophising about the impact pain will have on his functioning. Sullivan et al. (2006 **B**) suggest the role of catastrophising as a communication of the need for assistance from others. Therefore it can be seen as a passive coping strategy that Mr CP is using, possibly as he feels unable to self-manage his problem. Linton (2005 **A**) links catastrophising with distress, and both with the broader concept of anxiety. In that case, Mr CP's catastrophising could be interpreted as his anxiety about his current predicament, and not just a misconception.

Self-efficacy for self-management

Mr CP's previous management for his initial injury needs to be well documented, and the outcomes made clear. Interview should include determining his belief about the effectiveness of past treatments and finding out what his preferences are for managing his current problem.

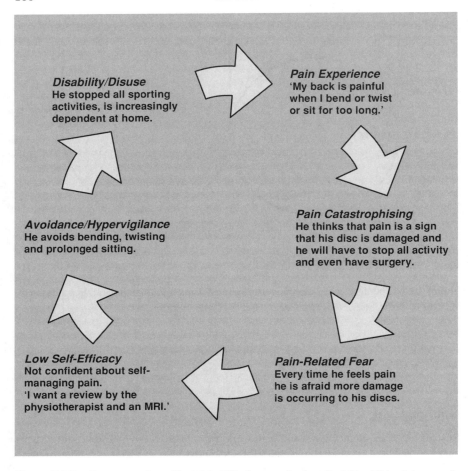

Figure 7.3.2a. Representation of how Mr CP's fear can lead to disability. This vicious cycle contributes to his pain experience and affects his ability to remain at work. The proposed treatment strategies will result in him being less avoidant and fearful of his back pain. His knowledge and problem solving skills will be enhanced and allow him to challenge his initial belief that he needs more treatment and investigations.

The passive treatments he describes have potentially contributed to a dependency on medical interventions and a disregard for self-management strategies. It would be of value to compare Mr CP's outcome expectations for treatments he has sought previously with his expectations for the results of managing his pain himself. If he is convinced that he has a damaged disc and that the only viable treatment is further surgery then he will be reluctant to engage in any other treatment, especially if he views it as potentially harmful. It is also important to ascertain how confident he is in performing self-management strategies. This is a self-efficacy belief and will be

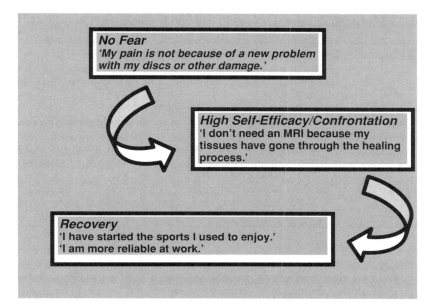

Figure 7.3.2b. Confrontation of fear and eventual recovery. The fact that Mr CP is questioning his unhelpful thoughts, is educated about his back pain, and is trying things out, means that he can challenge his unhelpful beliefs about his body being vulnerable.

influenced by his beliefs about the capabilities of his body, his belief about the nature of his condition, and his ability to do things despite the pain. Therefore assessment needs to address these issues, identifying any unhelpful beliefs about diagnosis and treatment and also any beliefs about his inability to perform the required tasks. Mr CP's behaviour appears to be associated with a low level of outcome expectation for self-management and a low level of self-efficacy for self-management, as reflected by his passive and avoidant coping style.

From yellow to blue to black flags

Research into condition-failed back surgery, where the patient has a poor outcome from surgery (Miller et al. 2005 **A**), reinforces the need to evaluate patients carefully, and especially for psychosocial factors known to influence outcome. Clinicians using manual therapies are also advised to screen for these factors, including using the Yellow Flag approach (Watson & Kendall 2000 **C**). Where the continuance of work or the return to work are important, assessment of blue and black flags are also indicated. These flags are associated specifically with occupational factors that may present as possible barriers to return to work (see Table 7.3.1).

Blue flags are concerned with perceptions related to work. Black flags relate to objective work characteristics. Blue flags are factors that are perceived by the worker

Table 7.3.1. Examples of the Yellow, Blue and Black Flags assessment approach to
Mr CP's case

Yellow Flags	Concern that pain is sign of disc damage.
	Expectation that surgery or manipulation will provide quick cure.
	Expectation that activities that cause an increase in pain should be avoided.
	Increasing dependence with domestic tasks.
Blue Flags	Concern that employer will not continue to be supportive due to his absence from work.
	Pressure to push himself harder in response to busyness at work.
	Expectation that workload will have to increase, although already very busy, if he is to climb career ladder.
Black Flags	Sickness absence management policy at work.
	Official work policy of reviewing performance if employees take more than 15 days' sick leave.

to be preventing them from returning to or continuing with work (Main & Burton 2000 **C**; Sowden 2006 **C**). Examples of what a worker might say include 'They expect me to work full time hours or not at all', and 'My employer doesn't believe me when I say I am in pain'. One identified blue flag in Mr CP's case is the concern that the initial support given by his boss may not continue. Also, he has indicated that he feels pressured at work and verbalises this through statements such as 'They ask me to take on more cases every time I am at work' and 'They expect me to stay back most days'. Due to the nature of his work, which includes a large amount of time in sitting, he feels that he is beginning to struggle to fulfil his duties. While he identifies a biological component for his inability to stay at work, saying 'Disc pain doesn't allow me to sit for long', there may be a component related to his expectation of the amount of work required of him, and what he perceives others expect of him in terms of work performance.

This is important to clarify during the assessment process. It may require some discussion in order for Mr CP to acknowledge the potential for multiple components, and engage in the pursuit and identification of psychosocial factors. In short, assessment can – perhaps should – be educational.

Black flags affect all workers equally. They include workplace policies and nationally established policies. Some examples of black flags are sickness policy; the role of occupational health in enforcing sickness policy; restricted duties; wage reimbursement rate (Main & Burton 2000 **C**). The place where Mr CP works has an official policy of reviewing an employee's performance after 15 days of sickness absence. This procedure is standard for any employee at his firm. This may result in him having a poor work record, being dismissed, or being in some other way penalised, for example missing a job promotion. This in turn may affect Mr CP's progress with rehabilitation and consequently set back his final goal of reducing work absence and improving quality of life.

A worse scenario for Mr CP is that his work absences become more frequent and of longer duration. Research identifies fear of re-injury, independent of pain severity, as a factor in long-term sick leave (Gheldof et al. 2005 **A**). Prognosis for those who are off work, or having difficulty returning to normal duties, for longer than 12 weeks are not encouraging (Airaksinen et al. 2005 **R**; Vendrig 1999 **A**). While the basis of this statement relies heavily on information from patients with workplace injuries, it highlights the importance of addressing Mr CP's pain management in the context of the work setting, to reduce any risks of long absence. The physiotherapist should involve themselves in this, but where workplace negotiations become complicated a workplace specialist should also be involved.

DIAGNOSIS

QUESTION 2

What is an appropriate label for Mr CP's low back pain?

IMAGING

As mentioned in Chapter 7.2, traditionally there has been an emphasis on tissue damage when considering an individual's pain and how to treat it. In particular, an individual's report of pain has been used as the main determinant in invasive treatments, including surgery and neurotomy. In Mr CP's case, he is hopeful that an MRI scan will help identify or diagnose a structural cause of his pain.

A number of studies since the mid 1990s have questioned the value of this. An investigation, using magnetic resonance imaging (MRI), into the structural integrity of intervertebral discs in asymptomatic subjects revealed that 56 % of the sample had disc lesions (Jensen et al. 1994 **A**). Further, four subjects were described as having disc protrusions, and one subject a disc extrusion. Yet these individuals were pain free. While there was no follow-up to see if these subjects developed pain later, the study demonstrates that despite the presence of identifiable tissue damage there is often no pain. One study which did follow up subjects (average follow-up was five years) found similar structural changes in asymptomatic subjects (Boos et al. 2000 **A**). While some back pain was reported in this group at follow-up, psychological factors and the nature of work were better predictors of medical consultation, than MRI findings. A more recent study demonstrated that there was no correlation between vertebral stress fracture or pars interarticularis defects, and pain or return to cricket (Millson et al. 2004 **A**). This included an example where pain persisted despite evidence of healing. Further, a recent study concluded that plain radiographs for low back pain add little value to therapeutic interventions, rarely detect serious pathology, and expose patients to radiation unnecessarily (van den Bosch et al. 2004). In any case, Ehrlich (2003 **R**) concludes that for disc pathology identified by imaging, invasive treatment commonly is ineffective. MRI for low back pain was found not to benefit treatment planning, and informing patients of the results may lead to greater worry (Modic et al. 2005).

Apart from patient preference, there is no support for Mr CP's request for an MRI scan (van Tulder et al. 2004 **R**), which is unlikely to diagnose the cause of the pain.

PASSIVE STRAIGHT LEG RAISE (PSLR)

Mr CP's PSLR was limited and asymmetrical. PSLR, and its variants, Lasegue's test and sign, has been considered a valuable predictor of disc herniation. However, Rebain, Baxter and McDonough (2002 **R**) undertook a systematic review of the use of PSLR in low back pain and found a need to clarify the role of psychosocial influences and muscle activity on the test. It was reported that psychological factors were not considered in any studies using the test. This greatly undermines its value as a diagnostic tool, given the potential for psychological factors to influence nervous system sensitivity and pain perception. There are elements of anxiety and distress in Mr CP's presentation which are likely to affect the specificity of the PSLR.

FAILED BACK SURGERY SYNDROME (FBSS) VERSUS NON-SPECIFIC LOW BACK PAIN (NSLBP)

While Mr CP's presentation could be considered NSLBP, the role of prior surgery in enhancing nervous system sensitivity may be better acknowledged with the FBSS label. Merksey and Bogduk (1994 **C**) insist in the IASP taxonomy that if surgery has been performed then it becomes the primary focus of the diagnostic label, rather than the pre-surgery diagnosis. So while Mr CP reports a disc pathology as the primary cause of symptoms (as well as 'facet joint'), FBSS or lumbar pain after failed surgery would be supported by current literature (Miller et al. 2005 **A**; Skaf et al. 2005 **A**).

TREATMENT

QUESTION 3

What is the best treatment for pain-related fear?

In response to Mr CP's biopsychosocial assessment, a brief problem list might be constructed, as in Table 7.3.2. Please note that this representation does not allow for the interactions of factors, or the potential impact of treatments on all aspects of the individual.

EVIDENCE FOR TREATMENT

The recent European Guidelines for acute NSLBP (van Tulder et al. 2004 **R**), chronic NSLBP (Airaksinen et al. 2005 **R**), and prevention in low back pain (Burton et al.

Table 7.3.2. Identified key treatment targets for physiotherapy from Mr CP's biopsychosocial assessment

Threatening Receptive Stimuli	Increased muscle tone in response to palpation examination. Sensory, proprioceptive and visual input interpreted as damaged or vulnerable lumbar spine.
Internal Beliefs and Emotions	Concerns and distress about disc injury and prognosis. Pain-related fear.
State and Structure of Nervous System	Sensitised due to above factors. Abnormal afferent and efferent activity due to reduced movement.

2004 **R**) provide evidence-based recommendations which will be considered in this section. Interestingly, there has been increasing support for the use of biopsychosocial interventions from initial contact in primary care (Grotle et al. 2005 **A**; Linton 2005 **A**; Pincus et al. 2002), which potentially blurs any distinction between 'acute' and 'chronic' pain management. As such, reference will be made to all three guidelines. It is important to recognise that those individuals who do not manage their acute pain well are likely to develop complex responses, including physical, neurological and psychological factors, and leading to reduced activity and distress, among other things. Those who do manage acute pain well, even if it persists to chronic pain, are unlikely to have the myriad of repercussions.

In terms of failed back surgery syndrome, there is some evidence that further surgery can help in select patients (Skaf et al. 2005). This will not be pursued in this section; instead it will be assumed that there is no identifiable tissue pathology.

BIOPSYCHOSOCIAL INTERVENTIONS

Physical therapy

There is some support for trialling manipulative therapy (including mobilisations) with Mr CP (Airaksinen et al. 2005 **R**; van Tulder et al. 2004 **R**). The physiotherapist should have a clear rationale for selecting this technique as it may lead to further undermining of a self-management approach.

Supervised exercise therapy is recommended by the 'chronic' guidelines, although no recommendations regarding specific exercises are made (Airaksinen et al. 2005 **R**). Given Mr CP's apparent concern about movement, supervised exercise may be valuable in providing support and encouragement. However, the context of his treatment (private clinic) would suggest other strategies, including a well-defined home exercise programme, may be more appropriate. The focus of this would be re-establishing normal range and movement, and therefore retraining the nervous system with regard to normal sensorimotor responses and safe limits.

Cognitive-behavioural interventions

Education and thought challenging

Education is recommended by the European guidelines (Airaksinen et al. 2005 **A**; Burton et al. 2004 **R**; van Tulder et al. 2004 **R**).

As stated before, the education process begins informally during assessment and should be part of the first meeting with Mr CP. This will help him to be informed for discussion of treatment goals. Open and collaborative education strategies are more likely to be empowering than a didactic approach.

Education has much support in the literature but the type of education is crucial. Education that targets patients' beliefs and emphasises behaviour change (Burton et al. 1999 **A**), aims to reduce the fear associated with pain (de Jong et al. 2005 **A**) and provides information about neurophysiology of pain (Moseley 2003 **A**, 2004 **A**), has better outcomes than education focusing on anatomical information about back structure, stability and back care. One session of appropriate education, including that by a physiotherapist, has repeatedly been shown to be effective (de Jong et al. 2005 **A**; Frost et al. 2004 **A**; Klaber Moffett et al. 2005 **A**; Moseley 2004 **A**). Notably in a number of these studies, researchers were specially trained in communication and use of cognitive-behavioural principles, or education was provided in a one-to-one context. Therefore a patient-centred approach to communication may optimise the value of education.

Presented in a non-threatening way, education would begin the essential process of thought-challenging and reconceptualisation. Mr CP should be allowed and encouraged to ask himself questions about his beliefs and behaviours related to pain. By challenging thoughts relating pain to disc damage, he will begin to consider challenging the behaviours associated with these thoughts, such as fear-avoidance (de Jong et al. 2005 **A**). Thought-challenging may need to be flagged as a useful coping strategy for moments of increased pain and distress.

It is also essential that when Mr CP challenges his beliefs he can consider a different explanation of his symptoms. In this way, he will understand how his nervous system can become sensitised and how mildly noxious stimuli or even non-noxious receptive information can thus be perceived as dangerous. Hopefully Mr CP will start some behavioural experiments, of his own accord, to confirm or deny this. If not, the physiotherapist may need to facilitate this; this is commonly achieved through a structured exercise programme or goal-setting task.

Graded exposure

While graded activity is mentioned as part of the multidisciplinary treatments in the guidelines, graded exposure is not specifically mentioned (Airaksinen et al. 2005 **A**; Burton et al. 2004 **R**; van Tulder et al. 2004 **R**).

Low back pain patients who score highly on the Fear-Avoidance Beliefs Questionnaire have been treated successfully using exposure therapy in the clinical and work or home settings (de Jong et al. 2005 A; Vlaeyen et al. 2002 A). Essentially this involves a hierarchy of activities or experiences – which might include visual imagery – that are ranked by the patient in the order of increasing fear or threat. The patient is then exposed to each level of the hierarchy, starting at the lowest. Common examples identified as fear-provoking activities in patients with pain are lifting a weight, bending forwards, being bumped by another person. Vlaeyen et al. (2002 A) determined that it is the graded exposure that is important in achieving change, and not just the graded activity that physiotherapists commonly prescribe.

On reflection, it is probably valuable for physiotherapists to reinterpret the outcomes of graded activity in order to recognise the potential for cognitive benefits, including supporting reconceptualisation, improving self-efficacy, and reducing fear. Also, by refining the approach to incorporate graded exposure strategies (such as a fear hierarchy), the intervention may be more effective. Importantly, graded exposure is usually performed in conjunction with clinical psychologists and caution must be taken to ensure the patient does not become more distressed. Physiotherapists need to be aware of their training needs and when it is appropriate to refer a person on to a clinical psychologist.

In Mr CP's case it would be expected that activities involving bending and twisting would feature highly on a fear hierarchy. Sitting for a prolonged period may feature in any graded exposure set in the workplace. He may need help to plan a hierarchical list of relevant fear-related activities. The physiotherapist may want to facilitate reflection of the process, particularly of how Mr CP feels after successfully meeting a challenge and of how success at the task relates to his predictions for the tasks. This may bring out evidence that Mr CP can use to challenge future unhelpful predictions of what might happen. Reflection will also help with designing a plan for the next exposure level on the hierarchy.

Goal-setting

Where a developing process is part of attainment of a treatment goal, structured goal-setting may be valuable. The patient may wish to do this independently or in collaboration with the physiotherapist (Klaber Moffett et al. 2006 C). However, despite evidence supporting the positive impact of goal-setting in physiotherapy (Arnetz et al. 2004 A), there is also evidence that physiotherapists do not do this well (Parry 2004 A) or fail to develop a shared level of understanding with the patient that would enable them to do this effectively (Daykin & Richardson 2004 A). To be effective in the collaboration, physiotherapists need to be able to provide guidance on evidence-based decision making. Such guidance has also been shown to be lacking (Bishop & Foster 2005 A).

Reflection

For Mr CP, goal-setting should incorporate reflection on the evaluation of past performances. Evidence of the levels of activity that his body is capable of, coping strategies that he has used previously to good effect, and his intimate knowledge of daily routines and home and work environments will be important features to consider. Reflections in preparation for goal-setting may provide useful insights into his method of coping in different contexts and situations.

Time-contingent, not pain-contingent

Use of quotas as a guide to activity is mentioned as part of graded activity/exercise in the guidelines (Airaksinen et al. 2005 **A**).

Mr CP obviously enjoys sports but is unable to participate at the level he desires. Also, his sitting tolerance is proving an issue at work. Improvement in his participation would benefit from a planned approach to pacing-up activity. Goal-setting should reflect a repetition- or time-contingent (such as a quota), not pain-contingent, approach to activity (Harding & Watson 2000 **C**). Correct baseline setting is essential. It should reflect a level of what Mr CP feels he can do regularly (Harding & Williams 1998 **C**). The starting baseline setting should emphasise a manageable level of activity, given that he will potentially be quite low skilled at managing his pain. Although Mr CP will be encouraged to do things despite pain, it is essential that he feels in control of it. That is, gritting teeth and pushing through the pain may only increase anxiety and raise nervous system sensitivity. Use of strategies such as relaxation, thought-challenging, and a planned gradual increase in activity level, will allow him to improve self-efficacy for self-management and perform activity despite pain, and without distress.

Structured planning to improve his sitting tolerance would be specifically beneficial, as this is obviously causing him some concern and may have a big impact on how he feels about and interacts with his workplace. There are time-contingent desensitisation strategies he could use to pace up his sitting tolerance (for example, using a timer), but he may find it inappropriate to implement these in some work situations (such as during a meeting with clients). Again, developing skill in a range of strategies will allow him to use what is effective in a given situation. This might even include self-talk such as 'I know this will make me sore but it is important I make a good impression'. However, this is unlikely to be helpful if it is associated with increased emotional distress.

Medium- and long-term goals

Once his confidence with the more immediate goals or initial levels of the fear hierarchy is increased, Mr CP may want to focus on what medium- and long-term goals he has to plan for. It may be important to set periods of time to allow reflection on progress, the skills already developed, and his achievements.

He may be worried that his absences from work will cause him to be viewed negatively in his performance appraisal. This could be true; however, he reported his boss was supportive. He could be in a position to work cooperatively with his boss, who may be able to help him stay at work for longer and work towards his promotion. He may want to initiate these types of negotiations in a meeting with his boss. This is potentially where a case manager or health care professional who is a workplace specialist could facilitate a workplace plan.

Flare-up or set-back plan

It is quite normal to experience fluctuations in the level of chronic pain. Mr CP may initially look to his physiotherapist to provide reassurance that this is normal. However, encouragement in self-management and self-reassurance is an important skill that he himself needs to develop.

It is also expected that there will be instances, extraordinary to these fluctuations common to persistent pain, when Mr CP will experience prolonged and increased levels of distress and pain. As part of managing these episodes, it is useful to have a well developed flare-up or set-back plan. Mr CP's flare-up plan might consist of a statement about remembering to do the strategies he has used effectively. It might include performing extra relaxation strategies or dedicating time to thought-challenging and reflection on helpful versus unhelpful coping strategies (for example catastrophising). It may also involve some activity management. Activity management usually does not require starting back at square one. It may mean not pacing-up for a period of time, consolidating the activity levels that he was managing before the flare up, or pacing-up at a slower rate.

Summary of treatment

Nociceptive triggers have not been identified and so were not specifically discussed in treatment. It is possible that something has not been identified and special attention should always be paid to following up on signs of serious pathology. However, it is important to recognise that pain may not have a nociceptive trigger and that the patient's emotions and beliefs and the sensitivity of the nervous system are equally important in the patient's report of pain.

By considering and addressing threatening receptive input, educating to modify beliefs and reduce distress, and normalising the experience of the nervous system, it is hoped that Mr CP will be more confident about managing painful episodes and even have a reduction in pain.

CONCLUDING REMARKS

Human pain is complex and demands complex solutions. The employment of cognitive-behavioural principles as outlined here directs management to be patient-centred, with an emphasis on self-management. Some patients may not be ready to participate in this way. However, by creating an open and non-threatening dialogue, the patients will be able to reflect on their beliefs and plan appropriate remediation of the problems they identify (Trede 2000 **A**).

While patients will look to physiotherapists to provide quality and expert opinion, it is just as important and much more empowering for patients to learn and employ the skills for self-management, rather than being told what to do. By taking this approach, the physiotherapist will create a treatment context that enables the patient to problem solve and safely explore their physical abilities. It gives the patient the opportunity to take control and explore the strategies that will best help them manage pain at home and in their work environment. The patient can take responsibility for some of the problem solving and therefore reduce the complexity of the challenge of pain.

ACKNOWLEDGEMENTS

To my family, especially WLS and KY. AW.

REFERENCES

Airaksinen O, Brox JI, Cedraschi C et al. (2005) European guidelines for the management of chronic non-specific low back pain. European Commission, Research Directorate General. http://www.backpaineurope.org Accessed 25 May 2006.

Arnetz JE, Bergstrom AK, Franzen Y, Nilsson H (2004) Active patient involvement in the establishment of physical therapy goals: effects on treatment outcome and quality of care. *Advances in Physiotherapy* **6**(2): 50–69.

Asghari A, Nicholas MK (2001) Pain self-efficacy beliefs and pain behaviour. *Pain* **94**(1): 85–100.

Asghari A, Nicholas MK (2006) Personality and pain-related beliefs/coping strategies: a prospective study. *Clinical Journal of Pain* **22**(1): 10–18.

Bishop A, Foster NE (2005) Do physical therapists in the United Kingdom recognise psychosocial factors in patients with acute low back pain? *Spine* **30**(11): 1316–1322.

Boos N, Semmer N, Elfering A, Schade V, Gal I, Zanetti M et al. (2000) Natural history of individuals with asymptomatic disc abnormalities in magnetic resonance imaging. *Spine* **12**: 1484–1492.

Burton AK, Tillotson KM, Main CJ, Hollis S (1995) Psychosocial predictors of outcome in acute and subchronic low back trouble. *Spine* **20**(6): 722–728.

Burton AK, Balagu F, Cardon G et al. (2004) European guidelines for prevention in low back pain. European Commission, Research Directorate General. http://www.backpaineurope. org Accessed 25 May 2006.

Burton AK, Waddell G, Tillotson KM, Summerton N (1999) Information and advice to patients with back pain can have a positive effect: a randomised controlled trial of a novel educational booklet in primary care. *Spine* **24**(23): 2484–2491.

Classen J, Cohen LG (2003) Practice-induced plasticity in the human motor cortex. In: Boniface S, Ziemann U (eds) *Plasticity in the Human Nervous System* Cambridge: Cambridge University Press, pp. 90–106.

Daykin AR, Richardson B (2004) Physiotherapists' pain beliefs: their influence on the management of patients with chronic low back pain. *Spine* **29**(7): 783–795.

de Jong JR, Vlaeyen JWS, Onghena P, Goossens MEJB, Geilen M, Mulder H (2005) Fear of movement/(re)injury in chronic low back pain. *Clinical Journal of Pain* **21**(1): 9–17.

Ehrlich GH (2003) Low back pain. *Bulletin of the World Health Organization* **81**: 671–676.

Flor H (2003) Cortical reorganisation and chronic pain: implications for rehabilitation. *Journal of Rehabilitation Medicine* **41**; Suppl.: 66–72.

Frost H, Lamb SE, Doll HA, Taffe Carver P, Stewart-Brown S (2004) Randomised controlled trial of physiotherapy compared to advice for low back pain. *British Medical Journal* http://www.bmj.com Accessed 26 May 2006.

Gheldof ELM, Vinck J, Vlaeyen JWS, Hidding A, Crombez G (2005) The differential role of pain, work characteristics and pain-related fear in explaining back pain and sick leave in occupational settings. *Pain* **113**: 71–81.

Grotle M, Brox JI, Veierod MB, Glomsrod B, Lonn JH, Vollestad NK (2005) Clinical course and prognostic factors in acute low back pain. *Spine* **30**(8): 976–982.

Harding V, Watson P (2000) Increasing activity and improving function in chronic pain management. *Physiotherapy* **86**(12): 619–629.

Harding V, Williams ACdeC (1998) Activities training: integrating behavioral and cognitive methods with physiotherapy in pain management. *Journal of Occupational Rehabilitation* **8**(1): 47–60.

Jellema P, van der Windt DAWN, van der Horst HE, Blankenstein AH, Bouter LM, Stalman WAB (2005) Why is treatment aimed at psychosocial factors not effective in patients with (sub)acute low back pain? *Pain* **118**: 350–359.

Jensen MC, Brant-Zawadzki MN, Obuchowski N, Modic MT, Malkasian D, Ross JS (1994) Magnetic resonance imaging of the lumbar spine in people without back pain. *New England Journal of Medicine* **331**(2): 69–73.

Kendall NAS, Linton SJ, Main CJ (1997) *Guide to Assessing Psychosocial Yellow Flags in Acute Low Back Pain: risk factors in long-term disability and work loss*. Wellington, New Zealand: Accident Rehabilitation and Compensation Insurance Corporation of New Zealand, and the National Health Committee, Ministry of Health.

Klaber Moffett J, Green A, Jackson D (2006) Words that help, words that harm. In: Gifford L (ed.) *Topical Issues in Pain 5* Falmouth: CNS Press.

Klaber Moffett J, Mannion AF (2005) What is the value of physical therapies for back pain? *Best Practice and Research* **19**(4): 623–638.

Klaber Moffett J, Jackson DA, Richmond S, Hahn S, Coulton S, Farrin A et al. (2005) Randomised trial of a brief physiotherapy intervention compared with usual physiotherapy for neck pain patients: outcomes and patients' preferences. *British Medical Journal* http://www.bmj.com Accessed 25 May 2006.

Linton SJ (2005) Do psychological factors increase the risk for back pain in the general population in both a cross-sectional and prospective analysis? *European Journal of Pain* **9**(4): 354–361.

Main CJ, Burton AK (2000) Economic and occupational influences on pain and disability. In: Main CJ, Spanswick CC (eds) *Pain Management: an interdisciplinary approach* London: Churchill Livingstone.

Merskey H, Bogduk N (1994) *Classification of Chronic Pain: descriptions of chronic pain syndromes and definition of chronic pain terms* (2 edn) Seattle: IASP Press.

Miller B, Gatchel RJ, Lou L, Stowell A, Robinson R, Polatin PB (2005) Interdisciplinary treatment of failed back surgery syndrome (FBSS): a comparison of FBSS and non-FBSS patients. *Pain Practice* **5**(3): 190–202.

Millson HB, Gray J, Stretch RA, Lambert MI (2004) Dissociation between back pain and bone stress reaction as measured by CT scan in young cricket fast bowlers. *British Journal of Sports Medicine* **38**(5): 586–591.

Modic MT, Obuchwski NA, Ross JS, Brant-Zawadzki MN, Grooff PN, Mazanec DJ et al. (2005) Acute low back pain and radiculopathy: MR imaging findings and their prognostic role and effect on outcome. *Radiology* **237**(2): 597–604.

Moseley GL (2003) Unravelling the barriers to reconceptualisation of the problem of chronic pain: the actual and perceived ability of patients and health professionals to understand neurophysiology. *Journal of Pain* **4**: 184–189.

Moseley GL (2004) Evidence for a direct relationship between cognitive and physical change during an education intervention in people with chronic low back pain. *European Journal of Pain* **8**: 39–45.

Moseley L (2006) Making sense of 'S1 mania'. In: Gifford L (ed.) *Topical Issues in Pain 5* Falmouth: CNS Press.

Parry R (2004) Communication during goal-setting in physiotherapy treatment sessions. *Clinical Rehabilitation* **18**: 668–682.

Peters ML, Vlaeyen JWS, Weber WEJ (2005) The joint contribution of physical pathology, pain-related fear and catastrophizing to chronic back pain disability. *Pain* **113**: 45–50.

Pincus T, Vlaeyen JWS, Kendall NAS, Von Korff NR, Kalauokalani DA, Reis S (2002) Cognitive-behavioral therapy and psychosocial factors in low back pain: directions for the future. *Spine* **27**(5): 133E–138E.

Rebain R, Baxter GD, McDonough S (2002) A systematic review of the passive straight leg raise test as a diagnostic aid for low back pain (1989 to 2000) *Spine* **27**(17): 388E–395E.

Robertson EM, Theoret H, Pascual-Leone A (2003) Skill learning. In: Boniface S, Ziemann U (eds) *Plasticity in the Human Nervous System* Cambridge: Cambridge University Press, pp. 107–134.

Sieben JM, Vlaeyen JWS, Portegijs PJM, Verbunt JA, van Riet-Rutgers S, Kester ADM et al. (2005) A longitudinal study on the predictive validity of the fear-avoidance model in low back pain. *Pain* **117**: 162–170.

Skaf G, Bouclaous C, Alaraj A, Chamoun R (2005) Clinical outcome of surgical treatment of failed back surgery syndrome. *Surgical Neurology* **64**: 483–489

Sowden G (2006) Vocational rehabilitation. In: Gifford L (ed.) *Topical Issues in Pain 6* Falmouth: CNS Press.

Storheim K, Brox JI, Holm I, Bo K (2005) Predictors of return to work in patients sick listed for sub-acute low back pain: a 12 month follow-up study. *Journal of Rehabilitation Medicine* **37**(6): 365–371.

Sullivan MJL, Thorn B, Rodgers W, Ward LC (2004) Path model of psychological antecedents to pain experience: experimental and clinical findings. *Clinical Journal of Pain* **20**: 164–173.

Sullivan MJL, Martel MO, Tripp DA, Savard A, Crombez G (2006) Catastrophic thinking and heightened perception of pain in others. *Pain* **123**: 37–44.

Swinkels-Meewisse IEJ, Roelofs J, Verbeek ALM, Ostendorp RAB, Vlaeyen JWS (2006) Fear-avoidance beliefs, disability, and participation in workers and nonworkers with acute low back pain. *Clinical Journal of Pain* **22**: 45–54.

Trede FV (2000) Physiotherapists' approaches to low back pain education. *Physiotherapy* **86**(8): 427–453.

van den Bosch MAAJ, Hollingworth W, Kinmonth KL, Dixon AK (2004) Evidence against the use of lumbar spine radiography for low back pain. *Clinical Radiology* **59**: 69–76.

van Nieuwenhuyse A, Somville PR, Crombez G, Burdorf A, Verbeke G, Johannik K et al. (2006) The role of physical workload and pain related fear in the development of low back pain in young workers: evidence from the BelCoBack Study; results after one year of follow up. *Occupational & Environmental Medicine* **63**(1): 45–52.

van Tulder M, Becker A, Bekkering T et al. (2004) European guidelines for the management of acute non-specific low back pain. European Commission, Research Directorate General. http://www.backpaineurope.org Accessed 25 May 2006.

Vendrig AA (1999) Prognostic factors and treatment-related changes associated with return to work in the multimodal treatment of chronic back pain. *Journal of Behavioral Medicine* **22**(3): 217–232.

Vlaeyen JWS, Linton SJ (2000) Fear-avoidance and its consequences in chronic musculoskeletal pain: a state of the art. *Pain* **85**: 317–332.

Vlaeyen JWS, de Jong J, Geilen G, Heuts PHTG, van Breukelen G (2002) The treatment of fear of movement/(re)injury in chronic low back pain: further evidence on the effectiveness of exposure in vivo. *Clinical Journal of Pain* **18**: 251–261.

Von Korff M, Balderson BHK, Saunders K, Miglioretti DL, Lin EHB, Berry S et al. (2005) A trial of an activating intervention for chronic back pain in primary care and physical therapy settings. *Pain* **113**: 323–330.

Watson P, Kendall N (2000) Assessing psychosocial yellow flags. In: Gifford L (ed.) *Topical Issues in Pain 2* Falmouth: CNS Press.

V Musculoskeletal

8 Evidence for Exercise and Self-Management Interventions for Lower Limb Osteoarthritis

NICOLA WALSH

CASE REPORT

BACKGROUND

Mrs S is a 62 year old female who lives with her husband in a semi-detached house with a large garden. They both retired two years ago, and now lead a relatively sedentary lifestyle, although they enjoy gardening and looking after their three grandchildren, who live locally. Mrs S has been experiencing intermittent pain in her right knee for approximately seven years, but has noticed a gradual increase in intensity over the last 18 months, including occasional discomfort in her left hip and knee; she is otherwise fit and well. She is now using her car more, as walking for more than twenty minutes aggravates the pain in her knees. She also reports stiffness in her hip and knee joints on waking, which resolves within 10–15 minutes of rising.

PREVIOUS MANAGEMENT

Mrs S has consulted her general practitioner (GP) several times for this problem, and was originally given paracetamol for pain relief and advised to lose weight; following a further GP consultation, she was also prescribed a non-steroidal anti-inflammatory drug. She has since been referred to physiotherapy, treated with acupuncture and given quadriceps exercises to perform at home on a daily basis. Although Mrs S initially noticed some relief in her pain following acupuncture, her pain has returned and is gradually worsening. Adherence to her exercise schedule has diminished, as she noticed minimal improvement in her symptoms and found the exercises boring, so stopped after one month. She has now been re-referred to physiotherapy.

CURRENT MEDICATION

- Paracetamol (2 twice daily).
- Glucosamine Sulphate (1000 mg daily).
- Rofecoxib (stopped medication 6 months ago).

Recent Advances in Physiotherapy. Edited by C. Partridge
© 2007 John Wiley & Sons, Ltd

MAIN DIAGNOSIS

- Primary osteoarthritis of the right knee (Kellgren-Lawrence Grade 2).
- Left hip and knee X-ray – NAD.

OBJECTIVE FINDINGS

- No structural deformity or instability of lower limb joints.
- Overweight (patient reports 7 kg weight increase over last 2 $^1/_2$ years).
- Normal gait pattern (c/o discomfort in right knee when standing from chair).
- Reduced quadriceps strength on right side.
- Decreased flexion in left and right knees (c/o end range stiffness).
- Full range of movement in both hips (c/o anterior 'tightness' on extension of left hip).

BASELINE ASSESSMENT OUTCOME MEASURES

- WOMAC = 19.
- Aggregate Functional Performance Time (AFPT) = 47.9 s.
- Self-efficacy for exercise = 63.

AGREED PROBLEM LIST

- Walking distance reduced due to discomfort.
- Muscle weakness in right leg.
- Stiffness in both knee joints.
- Tightness around left hip.
- Recent weight gain.

AIMS OF TREATMENT

- Increase comfortable walking distance.
- Improve lower limb function.
- Reduce pain.
- Provide patient with strategies to self-manage condition and encourage exercise adherence in the long-term.
- Encourage weight-loss.

TREATMENT PLAN

- Lower limb exercise and self-management OA class.
- Home exercise programme.

SIX WEEKS POST-INTERVENTION ASSESSMENT
OUTCOME MEASURES

WOMAC = 17.
AFPT = 38.4 s.
Self-efficacy = 73.

SIX MONTHS POST-INTERVENTION ASSESSMENT
OUTCOME MEASURES

WOMAC = 10.
AFPT = 36.1 s.
Self-efficacy = 69.

18 MONTHS POST-INTERVENTION ASSESSMENT
OUTCOME MEASURES

WOMAC = 14.
AFPT = 38.7 s.
Self-efficacy = 70.

INTRODUCTION

Osteoarthritis (OA) is the most common cause of pain, disability and functional impairment in the over-50 population, and increases in prevalence with age (Roddy et al. 2004 **C**). It is estimated that between 10 and 25 % of the post-retirement population experiences OA symptoms in the hip or knee joints alone (Petersson & Croft 1996 **B**), and there is a likelihood that these figures will increase as the size and longevity of this societal group expands. In addition, many people (recent figures suggest up to 8.5 million people in the UK) report symptoms of chronic joint pain with no formal diagnosis of OA, so the condition is even more prevalent than the figures suggest (Arthritis Care 2004 **B**).

The World Health Organisation (1997 **C**) cites OA as the fourth most prevalent disease amongst women in the developed world, and the eighth amongst males, establishing it as a considerable concern and burden to individuals, society and world health care systems. Economically, OA places huge financial demands on government and public spending. It is estimated that an annual societal cost of approximately £5.5 billion is incurred as a result of OA and chronic joint pain, including such factors as drug prescriptions; primary and secondary care conservative and surgical interventions; and lost revenue due to absence from work (Arthritis Research Campaign 2004 **B**; Hoffman et al. 1996 **B**; March & Bachmeier 1997 **B**). It is probable that there are considerable hidden costs that would escalate this figure further, for example, inclusion of iatrogenic pathology, and unpaid social support from family and friends (Leardini et al. 2004 **B**; Solomon et al. 2003 **B**).

The majority of patients with OA are managed within the community or primary care setting, often by their GP with simple analgesics and non-steroidal anti-inflammatory drugs (NSAID) (Peat et al. 2001 **R**), with less than 5 % progressing onto surgical intervention in secondary care (Walker-Bone et al. 2000 **R**). Although these medications relieve pain and have some impact on function (Superio-Cabuslay et al. 1996 **A**), they fail to address the underlying physical dysfunction in muscles and joints. In addition, drug interventions are costly (Leardini et al. 2004 **B**; Solomon et al. 2003 **B**) and often unpopular with patients (Chard et al. 2000 **R**; Tallon et al. 2000 **B**), and, as recent evidence suggests, prolonged NSAID use associated with co-morbid conditions common in older people, can induce serious or life-threatening side-effects (Hippisley-Cox & Coupland 2005 **B**; Hippisley-Cox et al. 2005 **B**). This evidence has resulted in withdrawal of several ubiquitously prescribed drugs from the market (Medicines and Healthcare products Regulatory Agency 2005 **C**).

Non-pharmacological treatment options still remain the cornerstone of primary care management, and international evidence-based guidelines strongly endorse the early use of exercise, weight loss and self-management/educational interventions for OA (American College of Rheumatology 2000 **C**; Jordan et al. 2003 **R**; Roddy et al. 2004 **C**).

Physiotherapy should underpin primary care management strategies, but considered and selective use of particular treatment modalities, and careful consideration of evidence of effectiveness, are essential if physiotherapy departments, which experience considerable time and financial constraints, are to adequately meet the demands of the increasing OA population.

In this chapter, I will briefly outline the pathological processes of OA in order to justify treatment selection, and then present contemporary best evidence to support physiotherapeutic interventions for the condition, in relation to clinical questions.

PATHOLOGY AND PRESENTATION

OA affects the synovial joint units of the musculoskeletal system, resulting in pain, sensorimotor dysfunction, decreased range of movement, and later stage deformity. Although pathological changes present in the hands and spine, OA is primarily burdensome for the individual and places the most extreme demands on health services when found in the hip and knee joints (Picavet & Hazes 2003 **B**).

OA is categorised into a primary and a secondary disorder, with the latter emanating from previous injury or biomechanical insult to the joint (Brandt et al. 1986 **C**). Causes of primary OA are more elusive, and although contemporary research attempts to identify genetic disorder and predisposition (Uitterlinden et al. 2000 **B**) or muscle dysfunction resulting in inadequate joint protection (Hurley 1999 **C**), the precise mechanism remains unknown. It should however be considered an active joint process, rather than simple 'wear and tear' as a result of aging (which it was traditionally held to be; this is now considered outdated and inappropriate).

The cartilage is predominantly vulnerable in OA, with disease processes resulting in softening, flaking, erosion, and ultimate disintegration of the collagenous matrix (Sandy & Verscahren 2001 **B**). However, articular cartilage is aneural, so cannot be responsible for the pain experienced in the early stages of the disease (Felson 2005 **B**). Concomitant changes in the bone, leading to sclerotic lesions and periarticular osteophytic formation, may indeed have a greater impact on pain levels and contribute to the malalignment deformities seen in the latter stages of the disease (Felson et al. 2003 **B**).

QUESTION 1

Which physiotherapy management strategies are beneficial for lower limb OA?

From a physiotherapy perspective, the changes in muscles are of primary interest, as they are rendered weak and susceptible to premature fatigue, but are plastic and therefore responsive to active intervention (Fisher & Prendergast 1997 **B**). Whether weakness results from arthrogenous muscle inhibition (Hurley & Newham 1993 **B**) or disuse atrophy associated with fear avoidance (Dekker et al. 1992 **R**), there is evidence implicating muscle tissue in the disease process.

The predominant complaints among patients with OA are of pain and loss of function (Peat et al. 2001 **R**), and as such, management strategies must focus on both the underlying physical and psychosocial dysfunctions that result in these problems.

Traditional understanding and physical treatments of OA were based on the premise that pathological changes impaired normal mechanical joint function, giving rise to pain and disability, and that those interventions which corrected such dysfunctions would ameliorate patient symptoms (Hurley & Newham 1993 **B**). This approach is embedded within the biomedical model of ill health and pain perception (Keefe et al. 2003 **C**), and provides a feasible explanation for the role of sensorimotor dysfunction in disease pathogenesis (Hurley 1999 **C**). However, it fails to account for the individual's comprehension and beliefs regarding their condition, emotions, previous experience, and a variety of other psychological sequelae that impact upon pain responses (Turk 1996 **C**).

A wider biopsychosocial model of ill health integrates the underlying pathology and physical dysfunction with the complex internal traits and external factors that influence pain perception, disease impact, and treatment response (Hurley et al. 2003 **B**; Jones et al. 2002 **B**). Indeed, it may also contribute to our understanding of the frequently cited incongruity between severity of radiographic OA and patient reported symptoms (Creamer et al. 1999 **A**), and disparity in success of, and response and adherence to, treatments (Hurley et al. 2003 **B**).

Mrs S had already undergone a course of physiotherapy involving acupuncture and a quadriceps home exercise regimen, with the former eliciting some short-term benefit, and the latter producing no self-reported improvements. Treatment strategies such as acupuncture, electrotherapy and manual therapy, each of which has a poor evidence-base in OA (Hurley & Walsh 2001 **R**) and may encourage patient reliance and passivity, coupled with advice and instruction to exercise at home, are a common approach to patient management (Walsh & Hurley 2005 **B**). However, effective treatment for OA involves active strategies that are patient controlled, encourage independence and long-term adherence, and have a strong evidence-base (Clarke 1999 **R**).

Therefore, the chosen treatment approach for Mrs S was an exercise and self-management class, integrating lower limb exercise with educational sessions that utilised active coping strategies for pain control, and discussed activity planning and overcoming barriers to exercise. The proposed aims of treatment were to encourage increased function and activity levels, improve but not eradicate pain, and promote long-term exercise adherence (full programme details can be found at www.kcl.ac.uk/gppc/escape).

QUESTION 2

Which type of exercise is most beneficial for lower limb OA?

Mrs S presented with reduced quadriceps strength bilaterally, with associated decreased range of knee joint movement, in addition to full range at her hips, but subjective joint 'tightness'. No structural malalignment or ligamentous laxity was noted. Walking was her predominant functional problem.

The evidence for the role of exercise in rehabilitation of OA is compelling, particularly for the knee joint, for which the majority of contemporary research has been undertaken. Although exercise is simple, accessible and cost-effective, careful consideration should be afforded to specificity, type and intensity, depending on local biomechanical factors (Sharma 2003 **B**). The documented benefits of exercise for hip OA are based on a very small number of underpowered studies and expert consensus opinion (Roddy et al. 2004 **C**), and the assumption that hip and knee joints affected by OA will respond in a similar manner. It is possible (although unlikely) that this is an inappropriate supposition, as latterly it has been speculated that hip OA and knee OA are site-specific sub-sets of the condition (Dennison & Cooper 2003 **C**). A recent meta-analysis of therapeutic exercise (strengthening and aerobic) for hip and knee OA demonstrated a combined effect size of 0.46 (95 % CI 0.35, 0.57) for improvements in self-reported pain, and an effect of 0.33 (95 % CI 0.23, 0.43) for self-reported physical function (Fransen et al. 2002 **A**). Although these effects may be deemed moderate, benefits may in fact be greater, as studies that used active control groups were included, which could dilute treatment effects. In addition, there was considerable heterogeneity within the studies – group and individual format, and aerobic and strengthening exercises – therefore it is very difficult to ascertain precise benefits.

In a subsequent analysis with disaggregated data for knee OA, aerobic exercise resulted in an effect size of 0.52 (95 % CI 0.34, 0.7) for pain and 0.46 (95 % CI

0.25, 0.67) for self-reported disability; and strengthening exercise demonstrated a 0.32 (95 % CI 0.23, 0.42) effect for pain and 0.32 (95 % CI 0.23, 0.41) for disability (Roddy et al. 2005 **A**).

The exercise class undertaken by Mrs S consisted of a variety of lower limb (particularly quadriceps) sub-maximal strengthening exercises: knee and hip range of movement; balance; and aerobic activities, lasting for 35–40 minutes, twice weekly. She was also provided with a mutually agreed home exercise programme tailored to her specific goal of increasing walking distance.

STRENGTHENING EXERCISE

Strengthening exercises for knee OA particularly focus on the quadriceps muscle group, due to the selective weakness commonly noted in patients with both clinical and radiographic degenerative changes (Slemenda et al. 1997 **B**) and the clear association between muscle weakness and decreased function (McAlindon et al. 1993 **B**). The role of the quadriceps group in knee function is to provide movement, support and sensorimotor feedback, and assist in load attenuation on contact (Hurley 1999 **C**). Therefore, the implicit assumption has been that rehabilitation of this muscle group will ameliorate these roles, thus enhancing the protective function over the degenerate joint.

The provision of strengthening exercises for knee OA was previously considered a standard 'prescription' irrespective of stage, extent, or the local joint environment. However, a recent study has questioned the viability of this approach, specifically in patients with malaligned or ligamentously lax joints (Sharma et al. 2003 **B**). This study notes that increased quadriceps strength at baseline is associated with greater (radiographic) progression in tibiofemoral degeneration, irrespective of alignment alterations. The authors suggest therefore that increasing quadriceps strength may affect force distribution around the knee joint, particularly the medial compartment in varus, and lateral compartment in valgus alignments (Sharma 2003 **B**), and is not optimal for the heterogeneous OA population. But there is no strong *direct* evidence at present to suggest that strengthening regimens increase the susceptibility of malaligned joints, and further large cohort, longitudinal studies will be necessary to confirm these postulations.

Hip muscle strengthening may also contribute to improvements in knee joint function relating to dynamic pelvic stability and foot angulation during gait (Hurwitz et al. 2002 **B**). A study by Yamada et al. (2001 **B**) found an increase in hip adductor strength in the presence of knee OA, which increased with disease severity. The authors postulated a theory that increased adductor strength reduced knee adduction moments, and was therefore actively employed by patients to reduce varus forces. As with other biomechanical studies however, further work is required to determine applicability to clinical practice.

There is good evidence to suggest strengthening exercises, particularly of the quadriceps, have a beneficial effect on pain and function in knee OA (Pelland et al. 2004 **A**). Further work clarifying the long-term effects on differing sub-sets of OA is necessary, to determine whether increased strength alters biomechanical factors that can positively and negatively impact on disease progression.

AEROBIC EXERCISE

Aerobic activity, particularly walking, is commonly incorporated into rehabilitation programmes for knee and to a lesser extent hip OA, both for its joint specific efficacy and its generic cardiovascular benefits (Bennell & Hinman 2005 **R**). A variety of exercises, including supervised walking (Kovar et al. 1992 **A**) and pedometer driven regimens (Talbot et al. 2003 **A**), have demonstrated benefits to pain and function. A recent meta-analysis reviewing the effects of aerobic activities (walking, jogging in water, T'ai Chi) in OA found clinically significant benefits for pain, and concluded aerobic activity was particularly effective for long-term functional outcomes (Brosseau et al. 2004 **A**). This suggests that adherence to exercise regimens is enhanced with activities that are more enjoyable and functionally orientated.

Walking is frequently encouraged in patients with lower limb OA, but physiotherapists should consider the manner in which it is prescribed, and indeed the desired functional outcome. Improving walking time and distance is appropriate, whereas increasing walking speed may be a detrimental aim for many patients. The evidence suggests that free speed walking is most appropriate, whereas higher rates may induce inappropriate forces around the knee, specifically in the presence of malalignment or reduced joint position sense (Hewett et al. 1998 **B**). Free speed walking in patients with hip OA induces minimal increase in contact pressures at the articulating surfaces, and considerably less than that caused by isometric muscle contraction or single-leg standing (Tackson et al. 1997 **B**). Further reductions in impact loading and joint reaction forces can be achieved through provision of shock absorbing insoles (Brouwer et al. 2005 **A**) or walking aids (Mendelson et al. 1998 **B**).

EXERCISE SUMMARY

It is probable that a combination of strengthening and aerobic exercise regimens is most beneficial for lower limb OA, but there is an increasing awareness of the need for specific biomechanical considerations for identified sub-groups. It is unlikely that a homogenous set of exercises will produce maximum benefit in a heterogeneous population, although this is unfortunately inherent in most research protocols and possibly in clinical practice. Alternative strategies that include specific motor-patterning or perturbation training may also warrant further research, particularly in the presence of functional instability.

In addition to the disparity in exercise type, there is considerable variation in intensity and duration of exercise trials. Many studies have impracticable and unreasonable intervention times (Messier et al. 2004 **A**; van Baar et al. 1998 **A**), which fail to translate into clinical practice, and may be unmanageable for patients in the longerterm. An exercise regimen that has sufficient time to allow patients to experience the benefits of exercise, to develop self-efficacy and confidence in their ability to exercise, whilst remaining clinically implementable, is preferable. Guidelines regarding exercise participation for general health and well-being suggest at least 30 minutes of light to moderate activity per day on most days of the week (Department of Health 2004 **C**). This figure may be cumulative, for example three short bouts of ten minutes, which is beneficial for many people with lower limb OA, who frequently find prolonged activity uncomfortable.

QUESTION 3

Can self-management programmes improve pain and function in patients with lower limb OA?

Possibly the most challenging aspect of exercise regimens is sustaining long-term adherence amongst patients once professionally supervised sessions have ceased. There is consistent research evidence to suggest that patients fail to comply with prescribed drug regimens (Haynes et al. 1996 **A**), and further support from the exercise literature to imply this is a generic problem (Ettinger et al. 1997 **A**; O'Reilly et al. 1999 **A**). Although this was traditionally considered a failure to accept and follow advice, contemporary thought suggests that adherence is in fact a complex, conscious reasoned process depending on preferences, beliefs, understanding, and experiences (Adams et al. 1997 **B**; Britten 1996 **C**). As such, researchers and clinicians should be mindful of the considerable psychosocial issues that influence patient decisions regarding treatments, and develop interventions that consider decision making and provide strategies to overcome adherence difficulties. In addition, establishing a habitual exercise behaviour is predictive of continued participation, and should therefore constitute an integral part of the management process (Rejeski et al. 1997 **A**).

This is the context in which self-management education programmes have been developed, to enable patients with chronic pain to acquire skills necessary to live functionally active lives. Prior to each exercise class, Mrs S participated in approximately 25–30 minutes' group discussion and self-management sessions led by a physiotherapist. The programme was based on self-efficacy theories, similar to the Arthritis Self-Management Programme, and consisted of sessions on: exercise action plans; management of diet and analgesia; ice, heat and relaxation as alternative methods of pain control; mood and pain perception; and overcoming barriers to exercise. Each session was supplemented with written information, which the patient was encouraged to file and refer to in the future (Hurley & Walsh 2005 **C**).

Patient education and self-management interventions are generally ill-defined terms that constitute an array of programmes for OA. Their purpose is to provide a patient with the skills and confidence to live a 'normal' life with their condition (Lorig 2003 **C**). Traditional physiotherapy approaches to patient education consisted of information delivery (generally regarding the pathological processes and beneficial effects of exercise) with minimal consideration of individual patient concerns, disease perceptions, condition beliefs and lifestyle changes. However, contemporary research suggests that it is necessary to engage patients in a rehabilitation partnership with the professional, and successful programmes need to build on self-efficacy and teaching the patient how to manage their problem (Lorig 2003 **C**; Rejeski et al. 1998 **B**).

Much of the research regarding multiple component education regimens has derived from the six week Arthritis Self-Management Programme (ASMP) developed by Lorig and Holman (1993 **R**). Originally designed as a lay-led programme in the USA, ASMP comprises a two hour session weekly for six weeks, delivered in a community setting to groups of 10–15 participants. Content includes activity planning, activity-rest cycling, relaxation techniques, and managing setbacks in progress (Lorig & Fries 1995 **C**). Studies of this programme in the UK demonstrate significant improvements in patients' health beliefs, implementation of exercise and healthy eating lifestyles, and psychological well-being (Barlow et al. 1999 **A**, 2000 **A**).

Whilst exercise demonstrates moderate effects on pain and function in OA, self-management interventions induce only small effect sizes. A meta-analysis of self-management education programmes for OA and rheumatoid arthritis (RA) produced pooled effect sizes of 0.12 for pain and 0.07 for disability (Warsi et al. 2003 **A**). The included studies were heterogeneous, both in content (self-efficacy and cognitive behavioural therapy approaches) and diagnosis (OA and RA), therefore beneficial effects on a particular sub-group with a specific approach may have been substantially diluted.

Mrs S had previously attended physiotherapy and was provided with knee exercises, which she stopped doing as she found them tedious and experienced little benefit. At the start of this period of rehabilitation, her self-efficacy for exercise score was moderate, but she expressed concern regarding exercise and activity, as walking in particular increased her knee pain and induced some hip discomfort. She had become less active as she thought this might be detrimental to her joint condition.

SELF-EFFICACY AND OSTEOARTHRITIS

The theory of self-efficacy postulated by Bandura (1977 **B**) considers an individual's beliefs regarding their ability to achieve personal goals and objectives, based on any previous experience of the task, the perceived benefits of the outcome, and their mastery of the necessary skills. In relation to OA, this translates into the capacity to self-manage or control various facets of the disease process, including functional capabilities, pain, and mood (Barlow 2001 **B**). There is a close relationship between self-efficacy, control, and helplessness. Although helplessness results in a consistent and general belief of diminished control, self-efficacy is task-specific and so can vary greatly within an individual (Hurley et al. 2003 **B**). As such it is necessary to target particular activities in order to maximise confidence and belief in performance abilities, allowing patients to experience the tangible, meaningful benefits of activity.

Previous research has established the important role of self-efficacy in OA. One study identified self-efficacy as an independent predictor of activity restriction in patients with knee OA (Rejeski 1996 **A**), while further research established that performance-related self-efficacy prospectively relates to functional decline in subjects with knee pain (Rejeski 2001 **B**). A recent study (n = 316) revealed that exercise, and exercise combined with diet interventions, increased self-efficacy for walking in subjects with knee OA, although only the combined intervention group reported an increase in self-efficacy for stair-climbing (Focht et al. 2005 **A**). These authors postulate that subjects require the intensity of the combined intervention in order to influence beliefs regarding their ability to undertake more demanding activities.

The complex interactions between self-efficacy and other psychosocial traits have a significant influence on an individual's belief regarding their ability to exercise, and on the health beliefs they attach to activity. Of particular importance is their understanding regarding the course and prognosis of their disease, as this can significantly impact on efficacy of interventions (Main & Watson 2002 **B**). Inappropriate health beliefs and anxiety can lead to fear avoidance and further joint degeneration, whereas correction of this behaviour can reduce depression and catastrophising, and encourage activity participation (Keefe et al. 1996 **C**).

FEAR AVOIDANCE

A commonly held belief amongst many OA patients is that the disease is an inevitable consequence of aging, with a relentless progression of joint degeneration that is exacerbated by activity. Consequently, erroneous beliefs create associations between exercise and harm – fear avoidance, a behaviour that results in further joint symptoms (Dekker 1992 **R**) (see Figure 8.1).

It is therefore a requirement of rehabilitation regimens that they challenge a patient's beliefs regarding their disease, not from a purely theoretical standpoint, but

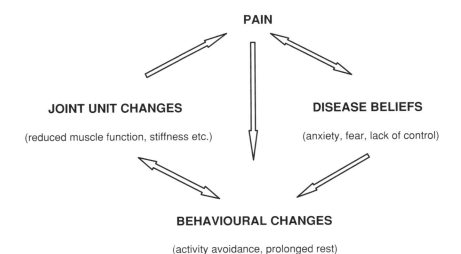

Figure 8.1. Interactions between pathology, health beliefs and behaviour in OA (adapted from Hurley et al. 2003 **B**).

by allowing them to experience the benefits of exercise and activity, increasing confidence and reinstating active coping mechanisms (Hurley et al. 2003 **B**). Integrated interventions that consist of patient self-management education and a participative exercise element are becoming increasingly prevalent.

QUESTION 4

Do combined exercise and self-management interventions improve pain and functional outcomes in lower limb OA?

Rehabilitation programmes that combine exercise regimens with patient education, self-management and coping strategies should maximise the benefits from both physical and educational approaches, and improve long-term adherence to activity and overall management of lower limb osteoarthritis.

The evidence suggests that combined interventions reduce pain (effect size 0.44; 95 % CI 0.70, 0.17) and increase function (effect size 0.27; 95 % CI 0.53, 0.002) in patients with lower limb OA (Walsh et al. 2006 **A**). However, these findings are based on a relatively small number of heterogeneous studies whose clinical applicability and practicability should be considered when judging the clinical implications.

Limited physiotherapy resources and an expanding elderly population render many of the integrated interventions proposed in research studies unmanageable, due to the time, logistic and financial constraints faced by clinical departments. The majority of programmes last for at least eight weeks (Fransen et al. 2001**A**; Hughes et al. 2004 **A**; Kuptniratsaikul et al. 2002 **A**), whilst one study continued physiotherapy and self-management input for 18 months (Messier et al. 2004). A study conducted by Hopman-Rock et al. (2000 **A**) of a clinically practicable intervention (2 hours a week for six weeks) combining group exercise and pain management sessions demonstrated clinically meaningful benefits six months post-intervention, and has since been implemented on a wider scale with equally successful outcomes (de Jong et al. 2004 **A**). Mrs S followed a twice weekly group programme for six weeks, each session lasting approximately one hour (Hurley & Walsh 2005 **C**). This regimen formed the intervention for a large RCT (n = 418), which showed beneficial effects on pain and function six months post-intervention (Hurley et al. 2005 **A**). Of note was the comparable effectiveness of group and individual interventions in this study, demonstrating both clinical- and cost-effectiveness. Combined exercise and self-management interventions are designed to promote long-term adherence to lifestyle changes and symptom control, and should therefore have lasting benefits. Most research studies provide limited follow-up however, and those that do extend their follow-up period report loss of short-term benefits without continued input (Quilty et al. 2003 **A**; van Baar et al. 2001 **A**). This raises questions regarding the long-term efficacy of combined interventions, within the current model of NHS care.

At present, following discharge, patients are left to manage their condition independently, with no planned follow-up. Clinically, this frequently leads to re-referrals, and as research demonstrates, results in reduced benefits of the initial intervention.

Innovative models of care may be necessary to improve the long-term efficacy of combined interventions, and manage the chronicity, changing nature, and concerns of patients with OA to greater effect.

A recent study looking at patient initiated consultations in rheumatoid arthritis (RA) demonstrated increased clinical- and cost-effectiveness in patients who sought medical/paramedical appointments when necessary, as compared with those who attended standard review appointments (Hewlett et al. 2005 A). Current management strategies do not accord the same importance to OA as to RA. A review of long-term medical and allied health professional management of the condition may be necessary to maximise the benefits derived from self-management interventions.

At six months post-intervention, Mrs S claimed to have taken fewer analgesics, and visited the GP less, so had improved the cost-efficiency of her management; disease specific and functional performance outcome measures had also remained stable. After a further 12 months however, her outcome measures had started to regress (although she still demonstrated improvements from baseline), indicating a decline in her functional condition.

QUESTION 5

For lower limb OA, can treatment efficacy be adequately measured?

Determining the effectiveness of any intervention is dependent on the outcome measures used, and the sensitivity of those measures to recognising improvements in the patient's condition (Chartered Society of Physiotherapy 2001 C). A survey of physiotherapy departments in the UK demonstrated that almost half of respondents failed to evaluate the outcome of their interventions (Walsh & Hurley 2005 B). Without evaluating what we do, we have no way of knowing whether we are safe, effective and efficient, and using our scant resources appropriately.

The integration of self-management interventions into standard physiotherapy treatments may require clinicians to familiarise themselves with measurement tools that elicit information from the psychosocial domain, rather than just the physical measures that are commonly utilised in physiotherapy (Walsh & Hurley 2005 B). Exercise self-efficacy measures (Bandura 1977 B) provide information on patients' beliefs regarding their ability to self-manage, while scales such as the Hospital Anxiety and Depression Scale (Zigmond & Snaith 1983 A) may provide an insight into the impact of the disease on patient mood status.

A disease-specific measure such as the Western Ontario and MacMaster Osteoarthritis (WOMAC) Index (Bellamy 1988 A) is frequently utilised in re-search studies, and provides information regarding pain, stiffness, and functional ability. It is easily administered and has demonstrated good validity and reliabil-ity. Used in conjunction with psychosocial measures and a functional outcome such as the Aggregate Functional Performance Time or Aggregate Locomotor Function (McCarthy & Oldham 2004 A), a comprehensive overview of the patient status and the efficacy of interventions can be established.

There are many disease-specific, generic, or patient generated outcome measures that are easy to apply, valid, and reliable, which would provide valuable information regarding practice. Clinical governance is reliant on outcome data to monitor success of treatment and appropriate use of resources, which makes outcome evaluation a requirement, not an option.

PRACTICE IMPLICATIONS

Many physiotherapy interventions lack sound scientific research to support their efficacy. Treatment regimens are often based on personal preference or experience, and weak empirical evidence, with efficacy determined on an individual basis in the clinical context.

For lower limb osteoarthritis however, there is a body of research of acceptable quality to support the use of exercise and self-management in improving pain and function. The greater challenge exists in integrating research findings into practice and encouraging clinicians to adopt these strategies as standard, in favour of other common modalities such as manual and electrotherapies, which may have significant placebo effects and be popular with patients and therapists, but have poor evidence of clinical efficacy.

In addition, maintaining the long-term benefits of combined interventions may require innovation in models of care, to ensure clinical- and cost-effective management, and further financial commitment to appropriate community facilities, in order to support therapeutic exercise for the older population.

Osteoarthritis is very common, and its prevalence will increase as the growing elderly population lives longer, placing further demands on an already financially stretched health service. As such, it is imperative that resources are utilised appropriately, to support interventions that demonstrate both clinical- and cost-effectiveness, and encourage patients to take responsibility for managing their own chronic condition.

CLINICAL BOTTOM LINE

- Good evidence exists to support the use of exercise and self-management strategies to treat lower limb osteoarthritis.
- Use of appropriate, functionally orientated outcome measures is essential for clinical governance and to determine the effectiveness of any intervention.
- Combined exercise and self-management programmes improve pain and function in the short-term, but clinical benefits are lost if patients do not adhere to lifestyle changes.
- Maintaining adherence to exercise and activity remains a challenge, and current service delivery and models of care may be insufficient long-term management strategies.

REFERENCES

Adams S, Pill R, Jones A (1997) Medication, chronic illness and identity: the perspective of people with asthma. *Social Science in Medicine* **45**: 189–201.

American College of Rheumatology (Subcommittee on Osteoarthritis Guidelines) (2000) Recommendation for the medical management of osteoarthritis of the hip and knee: 2000 update. *Arthritis and Rheumatism* **43**: 1905–1915.

Arthritis Care (2004) OA Nation. http://www.arthritiscare.org.uk/OANation Accessed December 2005.

Arthritis Research Campaign (2002) The Big Picture. http://www.arc.org.uk Accessed December 2005.

Bandura A (1977) Self-efficacy: toward a unifying theory of behavioral change. *Psychological Review* **84**: 191–215.

Barlow J (2001) How to use education as an intervention in osteoarthritis. *Clinical Rheumatology* **15**: 545–558.

Barlow JH, Turner A, Wright CC (2000) A randomized controlled study of the Arthritis Self Management Program in the UK. *Health Education Research: theory and practice* **15**: 665–680.

Barlow JH, Williams B, Wright C (1999) 'Instilling the strength to fight the pain and get on with life': learning to become an arthritis self-manager through an adult education program. *Health Education Research: theory and practice* **14**: 533–544.

Bellamy N, Buchanan WW, Goldsmith CH et al. (1988) Validation study of WOMAC: a health status instrument for measuring clinically important patient relevant outcomes to antirheumatic drug therapy in patients with osteoarthritis of the hip or knee. *Journal of Rheumatology* **15**: 1833–1840.

Bennell K, Hinman R (2005) Exercise as a treatment for osteoarthritis. *Current Opinion in Rheumatology* **17**: 634–640.

Brandt KD, Mankin HJ, Shulman LE (1986) Workshop on etiopathogenesis of osteoarthritis. *Journal of Rheumatology* **13**: 1126–1160.

Britten N (1996) Lay views of drugs and medicines: orthodox and unorthodox accounts. In: Williams SJ, Calnan M (eds) *Modern Medicine: lay perspectives and experiences* London: UCL Press, pp. 48–73.

Brosseau L, Pelland L, Wells G et al. (2004) Efficacy of aerobic exercises for osteoarthritis (part II): a meta-analysis. *Physical Therapy Reviews* **9**: 125–145.

Brouwer RW, Jakma TSC, Verhagen AP et al. (2005) Braces and orthoses for treating osteoarthritis of the knee joint: a systematic review. *Cochrane Library* **4** http://www.thecochranelibrary.com.

Chard J, Tallon D, Dieppe PA (2000) Epidemiology of research into interventions for the treatment of osteoarthritis of the knee joint. *Annals of the Rheumatic Diseases* **59**: 414–418.

Chartered Society of Physiotherapy (2001) *Outcome Measures. Report CLEF03* Chartered Society of Physiotherapy.

Clarke AK (1999) Effectiveness of rehabilitation in arthritis. *Clinical Rehabilitation* **13**(1): 51–62.

Creamer P, Lethbridge-Cejku M, Hochberg MC (1999) Determinants of pain severity in knee osteoarthritis: effect of demographic and psychosocial variables using 3 pain measures. *Journal of Rheumatology* **26**: 1785–1792.

de Jong ORW, Hopman-Rock M, Tak ECMP, Klazinga NS (2004) An implementation study of two evidence-based exercise and health education programmes for older adults with osteoarthritis of the knee and hip. *Health Education Research* **19**: 316–325.

Dekker J, Bott B, van der Woude LHV, Bijlsma JWJ (1992) Pain and disability in osteoarthritis: a review of biobehavioural mechanisms. *Journal of Behavioural Medicine* **15**: 189–214.

Dennison E, Cooper C (2003) The natural history and prognosis of osteoarthritis. In: Brandt K, Doherty M, Lohmander S (eds) *Osteoarthritis* (2 edn) Oxford: Oxford University Press, pp. 227–233.

Department of Health (2004) *At Least Five a Week: evidence on the impact of physical activity and its relationship to health. A report from the Chief Medical Officer.* Department of Health.

Ettinger WH, Burns R, Messier SP et al. (1997) A randomized trial comparing aerobic exercise and resistance exercise with a health education program in older adults with knee osteoarthritis. *Journal of the American Medical Association* **277**: 27–31.

Felson, DT (2005) The sources of pain in knee osteoarthritis. *Current Opinion in Rheumatology* **17**: 624–628.

Felson DT, McLaughlin S, Goggins J et al. (2003) Bone marrow edema and its relation to progression of knee osteoarthritis. *Annals of Internal Medicine* **139**: 330–336.

Fisher NM, Prendergast DR (1997) Reduced muscle function in patients with osteoarthritis. *Scandinavian Journal of Rehabilitation Medicine* **29**: 213–221.

Focht BC, Rejeski WJ, Ambrosius WT et al. (2005) Exercise, self-efficacy, and mobility performance in overweight and obese older adults with knee osteoarthritis. *Arthritis and Rheumatism* **53**: 659–665.

Fransen M, Crosbie J, Edmonds J (2001) Physical therapy is effective for patients with osteoarthritis of the knee: a randomised controlled trial. *Journal of Rheumatology* **28**: 156–164.

Fransen M, McConnell S, Bell M (2002) Therapeutic exercise for people with osteoarthritis of the hip or knee: a systematic review. *Journal of Rheumatology* **29**: 1737–1745.

Haynes RB, McKibbon KA, Kanani R (1998) Systematic review of randomised trials of interventions to assist patients to follow prescriptions for medications. *Lancet* **348**: 383–386.

Hewett T, Noyes F, Barber-Westin S, Heckmen T (1998) Decrease in knee joint pain and increase in function in patients with medial compartment arthrosis: a prospective analysis of valgus bracing. *Orthopaedics* **21**: 131–138.

Hewlett S, Kirwan J, Pollock J et al. (2005) Patient initiated outpatient follow up in rheumatoid arthritis: six year randomised controlled trial. *British Medical Journal* **330**: 171–175.

Hippisley-Cox J, Coupland C (2005) Risk of myocardial infarction in patients taking cyclo-oxygenase-2 inhibitors or conventional non-steroidal anti-inflammatory drugs: population based nested case-control analysis. *British Medical Journal* **330**: 1366–1369.

Hippisley-Cox J, Coupland C, Logan R (2005) Risk of adverse gastrointestinal outcomes in patients taking cyclo-oxygenase-2 inhibitors or conventional non-steroidal anti-inflammatory drugs: population based nested case-control analysis. *British Medical Journal* **331**: 1310–1316.

Hoffman C, Rice D, Sung H-Y (1996) Persons with chronic conditions. Their prevalence and costs. *Journal of the American Medical Association* **276**(18): 1473–1479.

Hopman-Rock M, Westhoff MA (2000) The effects of a health educational and exercise program for older adults with osteoarthritis of the hip or knee. *Journal of Rheumatology* **27**: 1947–54.

Hughes SL, Seymour RB, Campbell R et al. (2004) Impact of the fit and strong intervention on older adults with osteoarthritis. *The Gerontologist* **44**: 217–228.

Hurley MV (1999) The role of muscle weakness in the pathogenesis of osteoarthritis. *Rheumatic Disease Clinics of North America* **25**: 283–298.

Hurley MV, Mitchell HL, Walsh NE (2003) In osteoarthritis, the psychosocial benefits of exercise are as important as physiological improvements. *Exercise and Sports Science Reviews* **31**(3): 138–143.

Hurley MV, Newham DJ (1993) The influence of arthrogenous muscle inhibition on quadriceps rehabilitation of patients with early unilateral osteoarthritic knees. *British Journal of Rheumatology* **32**: 127–131.

Hurley M, Walsh N (2005) ESCAPE knee pain: a programme handbook. http://www.kcl.ac.uk/gppc/escape Accessed February 2006.

Hurley M, Walsh N (2001) Physical, functional and other non-pharmacological interventions for osteoarthritis. *Clinical Rheumatology* **15**(4): 569–581.

Hurley MV, Walsh NE, Mitchell HL et al. (2005) Enabling self-management and coping with arthritic pain through exercise (ESCAPE-knee pain): a cluster randomised trial of an integrated rehabilitation programme. *Arthritis and Rheumatism* **52**: 505S.

Hurwitz D, Ryals A, Case J et al. (2002) The knee adduction moment during gait in subjects with knee osteoarthritis is more closely correlated with static alignment than radiographic disease severity, toe out angle and pain. *Journal of Orthopaedic Research* **20**: 101–108.

Jones M, Edwards I, Gifford L (2002) Conceptual models for implementing biopsychosocial theory in clinical practice. *Manual Therapy* **7**: 2–9.

Jordan KM, Arden NK, Doherty M, Bannwarth B et al. (2003) EULAR Recommendations: an evidence based approach to the management of knee osteoarthritis: report of a task force of the Standing Committee for International Clinical Studies Including Therapeutic Trials (ESCISIT). *Annals of the Rheumatic Diseases* **62**(12): 1145–1155.

Keefe F, Aspnes A, Caldwell DS, Kashikar-Zuck S (2003) Coping strategies for the patient with osteoarthritis. In: Brandt K, Doherty M, Lohmander S (eds) *Osteoarthritis* (2 edn) Oxford: Oxford University Press, pp. 339–346.

Keefe FJ, Kashnikar-Zuck S, Opiteck E et al. (1996) Pain in arthritis and musculoskeletal disorders: the role of coping skills training and exercise interventions. *Journal of Orthopaedics and Sports Physical Therapy* **24**: 279–290.

Kovar PA, Allegrante JP, MacKenzie R et al. (1992) Supervised fitness walking in patients with osteoarthritis of the knee: a randomized controlled trial. *Annals of Internal Medicine* **116**: 529–534.

Kuptniratsaikul V, Tosayanonda O, Nilganuwong S et al. (2002) The efficacy of a muscle exercise program to improve functional performance of the knee in patients with osteoarthritis. *Journal of the Medical Association of Thailand* **85**: 33–40.

Leardini G, Salaffi F, Caporali R et al. (2004) Direct and indirect costs of osteoarthritis of the knee. *Clinical and Experimental Rheumatology* **22**: 699–706.

Lorig K (2003) Self-management education: more than a nice extra. *Medical Care* **41**: 699–701.

Lorig K, Fries J (1996) *The Arthritis Helpbook* Reading: Addison-Wesley.

Lorig K, Holman H (1993) Arthritis self-management studies: a twelve year review. *Health Education Quarterly* **20**: 17–28.

Main C, Watson P (2002) Psychological aspects of pain. *Manual Therapy* **4**: 203–215.

McAlindon TE, Cooper C, Kirwan JR et al. (1993) Determinants in disability in osteoarthritis of the knee. *Annals of the Rheumatic Diseases* **52**: 258–262.

March LM, Bachmeier CJM (1997) Economics of osteoarthritis: a global perspective. *Bailliere's Clinical Rheumatology* **11**: 817–834.

McCarthy CJ, Oldham JA (2004) The reliability, validity and responsiveness of an aggregate locomotor function (ALF) score in patients with osteoarthritis of the knee. *Rheumatology* **43**: 415–517.

Medicines and Healthcare products Regulatory Agency. Cox-2 inhibitors. http://www.mhra. gov.uk/home/idcplg?ldcService=SS_GET_PAGE&nodeld=227 Updated 13 December 2005. Accessed 3rd January 2006.

Mendelson S, Milgron C, Finestone A et al. (1998) Effect of cane use on tibial strain and strain rates. *American Journal of Physical Medicine and Rehabilitation* **11**: 333–338.

Messier SP, Loeser RF, Miler GD et al. (2004) Exercise and dietary weight loss in overweight and obese older adults with knee osteoarthritis: the Arthritis, Diet, and Activity Promotion Trial. *Arthritis and Rheumatism* **50**: 1501–1510.

O'Reilly SC, Muir KR, Doherty M (1999) The effectiveness of home exercise on pain and disability from osteoarthritis of the knee: a randomised controlled trial. *Annals of the Rheumatic Diseases* **58**: 15–19.

Peat G, Croft P, Hay E (2001) Clinical assessment of the osteoarthritis patient. *Best Practice and Research Clinical Rheumatology* **15**(4): 527–544.

Pelland L, Brosseau L, Wells G et al. (2004) Efficacy of strengthening exercises for osteoarthritis (part I): a meta-analysis. *Physical Therapy Reviews* **9**: 77–108.

Petersson I, Croft P (1996) Occurrence of osteoarthritis in the peripheral joints in European populations and outside Europe. *Annals of the Rheumatic Diseases* **55**: 659–664.

Picavet HSJ, Hazes JMW (2003) Prevalence of self-reported musculoskeletal diseases is high. *Annals of Rheumatic Disease* **62**: 644–650.

Quilty B, Tucker M, Campbell R, Dieppe P (2003) Physiotherapy, including quadriceps exercises and patellar taping, for knee osteoarthritis with predominant patello-femoral joint involvement: randomized controlled trial. *Journal of Rheumatology* **30**: 1311–1337.

Rejeski WJ, Miller ME, Foy C et al. (2001) Self-efficacy and the progression of functional limitations and self-reported disability in older adults with knee pain. *Journal of Gerontology Series B* **56**: 261–265.

Rejeski WJ, Ettinger WH, Martin K, Morgan T (1998) Treating disability in knee osteoarthritis with exercise therapy: A central role for self-efficacy and pain. *Arthritis Care and Research* **11**: 94–101.

Rejeski WJ, Brawley LR, Ettinger W et al. (1997) Compliance to exercise therapy in older participants with knee osteoarthritis: implications for treating disability. *Medicine and Science in Sports and Exercise* **29**: 977–985.

Rejeski WJ, Craven T, Ettinger WH et al. (1996) Self-efficacy and pain in disability with osteoarthritis of the knee. *Journal of Gerontology Series B* **51**: 24–29.

Roddy E, Zhang W, Doherty M (2005) Aerobic walking or strengthening exercise for osteo-arthritis of the knee? A systematic review. *Annals of Rheumatic Disease* **64**: 544–548.

Roddy E, Zhang W, Doherty M et al. (2004) Evidence-based recommendations for the role of exercise in the management of osteoarthritis of the hip or knee – the MOVE consensus. *Rheumatology* **44**(1): 67–73.

Sandy JD, Verscharen C (2001) Analysis of aggrecan in human knee cartilage and synovial fluid indicates that agrecanase ADAMTS activity is responsible for the catabolic turnover and loss of whole aggrecan whereas other protease activity is required for C-terminal processing in vivo. *Biochemistry Journal* **358**: 615–626.

Sharma L (2003) Examination of exercise effects on knee osteoarthritis outcomes: why should the local mechanical environment be considered? *Arthritis and Rheumatism* **49**(2): 255–260.

Sharma L, Dunlop DD, Cahue S et al. (2003) Quadriceps strength and osteoarthritis progression in malaligned and lax knees. *Annals of Internal Medicine* **138**: 613–619.

Slemenda C, Brandt KD, Heilman K et al. (1997) Quadriceps weakness and osteoarthritis of the knee. *Annals of Internal Medicine* **127**: 97–104.

Solomon DH, Glynn RJ, Bohn R et al. (2003) The hidden cost of non-selective nonsteroidal anti-inflammatory drugs in older patients. *Journal of Rheumatology* **30**: 792–798.

Superio-Cabuslay E, Ward MM, Lorig KR (1996) Patient education interventions in osteo-arthritis and rheumatoid arthritis: a meta-analytic comparison with non-steroidal anti inflammatory drug treatment. *Arthritis Care and Research* **9**: 292–301.

Tackson SJ, Krebs DE, Harris BA (1997) Acetabular pressures during hip arthritis exercises. *Arthritis Care and Research* **10**: 308–319.

Talbot LA, Gaines JM, Huynh TN, Metter EJ (2003) A home-based pedometer driven walking program to increase physical activity in osteoarthritis of the knee: a preliminary study. *Journal of the American Geriatric Society* **51**: 387–392.

Tallon D, Chard J, Dieppe P (2000) Exploring the priorities of patients with osteoarthritis of the knee. *Arthritis Care and Research* **13**: 312–319.

Turk DC (1996) Biopsychosocial perspectives on chronic pain. In: Gatchel RJ, Turk DC (eds) *Psychological Approaches to Pain Management: a practitioner's handbook* New York: Guildford Press, pp. 3–32.

Uitterlinden AG, Burger H, van Duijn CM et al. (2000) Adjacent genes, for COL2A1 and the vitamin D receptor, are associated with separate features of radiographic osteoarthritis of the knee. *Arthritis and Rheumatism* **43**: 1456–1464.

van Baar ME, Dekker J, Oostendorp RAB at al. (1998) The effectiveness of exercise therapy in patients with osteoarthritis of the hip or knee: a randomized clinical trial. *Journal of Rheumatology* **25**: 2432–2439.

van Baar ME, Oostendorp RAB, Bijl D et al. (2001) Effectiveness of exercise in patients with osteoarthritis of hip or knee: nine months' follow up. *Annals of Rheumatic Diseases* **60**: 1123–1130.

Walker-Bone K, Javaid K, Arden N, Cooper C (2000) The medical management of osteoarthritis. *British Medical Journal* **321**: 936–940.

Walsh NE, Hurley MV (2005) Management of knee osteoarthritis in physiotherapy out-patient departments in Great Britain and Northern Ireland. *Rheumatology* **44**; Suppl. 1: 145i.

Walsh NE, Hurley MV, Mitchell HL, Reeves BC (2005) The effects of combined exercise and self-management regimens on pain and function in patients with osteoarthritis of the hip and knee: a systematic review with meta-analysis. *Arthritis and Rheumatism* **52**: 717S.

Walsh NE, Mitchell HL, Reeves BC, Hurley MV (2006) Integrated exercise and self-management programmes in osteoarthritis of the hip and knee: a systematic review of effectiveness. *Physical Theory Reviews* **11**. In press.

Warsi A, La Valley MP, Wang PS et al. (2003) Arthritis self-management programs. A meta-analysis of the effect on pain and disability. *Arthritis and Rheumatism* **48**: 2207–2213.

World Health Organisation (1997) *The Global Burden of Disease* Geneva: World Health Organisation.

Yamada H, Koshino T, Sakai N et al. (2001) Hip adductor muscle strength in patients with knee osteoarthritis with varus deformed knee. *Clinical Orthopaedics* **60**: 612–618.

Zigmond AS, Snaith RP (1983) The hospital anxiety and depression scale. *Acta Psychiatrica Scandinavia* **67**: 61–70.

9 Using Evidence-Based Practice for Upper Extremity Musculoskeletal Disorders

JOY C. MACDERMID

INTRODUCTION

'Musculoskeletal (MSK) disorder' is a broad term encompassing a variety of disorders that affect the MSK system. In this chapter we focus on gradual onset upper extremity disorders (UED) and present two cases that represent common pathologic processes: tendinopathy and compressive neuropathy. The multifactorial nature of UED is becoming increasingly apparent as causation crosses psychological, physical, and environmental factors (Aaras et al. 2001 **A**; Baker et al. 1999 **A**; Bongers et al. 2002 **R**; de Jonge et al. 2000 **A**; Devereux et al. 2002 **A**; Feuerstein et al. 2000 **A**; Feuerstein et al. 2004 **A**; Himmelstein et al. 1995 **A**; Huang et al. 2003 **A**; Johansson & Rubenowitz 1994 **A**; Lundberg 2002 **B**; Novak & Mackinnon 2002 **A**; Warren 2001 **A**). This spectrum of disease and the multifactorial nature of its causation creates a profound barrier to accurate classification, which hampers progress on defining the epidemiology, causation, prognosis, and optimal management of MSK disorders (Van Eerd et al. 2003 **R**). Variation between studies in terms of spectrum of disease contributes to variability in results and conclusions and may, in part, explain some of the conflicting results in published literature. This has been mentioned as a limitation in numerous MSK studies. Clinicians who use evidence-based practice to optimise the quality of care for UED must be prepared to deal with resulting uncertainty.

The basic principles of evidence-based practice are covered elsewhere, but it is worthwhile considering how to apply these principles in UED (Sackett et al. 2000 **C**). The basic steps are as described in this series and in other chapters. However, defining an appropriate clinical question is challenging in UED because of the uncertainty around the cause, diagnosis, and severity of many of the problems. Clinical questions regarding UED can be derived using clinical experience and a patient-centered approach. First identify the diagnosis and/or impairments that are causing disability or limiting participation in meaningful life roles, and then use the patient's goals to derive clinical questions that have meaning to both patient and therapist. It is especially important to understand the theoretical and biological bases of these clinical questions

Recent Advances in Physiotherapy. Edited by C. Partridge
© 2007 John Wiley & Sons, Ltd

in order to deal with the uncertainty inherent in UED. In cases where clinical data is absent, it is important to have a strong biological and theoretical foundation for the treatment principles involved.

CASE REPORT I

Mr AE is a 46 year old male who works in an automobile manufacturing plant. During the initial interview, he stated that he has right elbow pain when he moves his arm. This pain sometimes goes down the arm and is usually 'achy' after activity or at the end of the day, but occasionally a 'sharper pain' occurs with certain movements. These include using tools at work, wringing out a wet towel, and carrying a heavy pot. On further questioning, Mr AE said he hurt the elbow 10 days ago while turning a 'tight bolt'. He noticed right lateral elbow pain following this task, but was able to continue his entire shift. He reported having had similar difficulty on other occasions, but these tended to resolve in several days. This time he noticed the discomfort continued over his entire shift. While the pain was better the next morning, it continued to be aggravated by many activities that required the use of his elbow or wrist on the following day, and persisted for a week. At that point, he went to his family physician, who prescribed naproxen (250 mg twice daily) and referred him to physiotherapy. Other than this problem, he is in good health. His DASH score is 30.

PERTINENT FINDINGS ON PHYSICAL EXAMINATION

- Palpation: mild tenderness at the lateral aspect of the right elbow.
- Joint Motion:
  - AROM:
    - Left elbow full (5° hyperextension to 145° flexion) and pain free.
    - Right elbow full (5° hyperextension to 143° flexion) and pain free.
    - Left and right superior radio-ulnar joint full (supination 90°, pronation 83°) and pain free.
  - PROM:
    - Left and right elbow full and pain free; normal end-feels.
  - Combined Movements: full extension with full pronation reproduces pain at elbow; other combined movements are full range and pain free.
- Static Muscle Testing:
  - Resisted elbow flexion and extension; pain free; normal power.
  - Resisted pronation and supination; pain free; normal power.
  - Resisted wrist extension (with elbow extended); painful; weak on right.
  - Resisted wrist flexion is pain free with elbow in all positions; normal power.
- Measured Strength:
  - Left: elbow extension 110 N; elbow flexion 134 N; maximum grip 34 kg; pain-free grip 34 kg.
  - Right: elbow extension 117 N; elbow flexion 143 N; maximum grip 33 kg; pain-free grip 21 kg.

- Stability: pain free and good stability on medial and lateral stress.
- Sensation: normal – light touch.

QUESTION 1

What is the best approach for successful conservative management of this patient's problem (tennis elbow)?

First, we used our knowledge of this condition to develop appropriate questions. Lateral epicondylosis (LE) is a common disorder in workers who perform repetitive forearm motions with gripping. Repetitive activity contributes to degenerative changes of the tendon, particularly at its insertion. Recently, the role of degenerative changes has been emphasised (Kraushaar & Nirschl 1999 **R**), although clinical experience tells us that a small subgroup of patients with acute cases of tennis elbow may fit more of an inflammatory model. Members of this subgroup are likely to be different (Wuori et al. 1998 **A**) in pathology, prognosis and, therefore, should be approached differently.

Our patient has a subacute or episodic aggravation of ongoing LE and likely has some degenerative changes. Our first question was general in nature. We wanted to identify a basic approach to management of lateral epicondylosis. We searched the literature and found two clinical practice guidelines (MacDermid 2004 **A**), but both were uninformative and low quality, so they were discarded. The next step was to search for a relevant systematic review and we found that there have been a number published that address the effectiveness of various treatments for lateral epicondylosis, creating a good starting point to get an overview of the most evidence-based approach.

In 1996, Assendelft et al. (1996 **R**) looked at the effectiveness of corticosteroid injections for lateral epicondylitis. This review found that at that time, no conclusive reports could be made on the effectiveness of the injections. This was due to the serious methodological flaws found in the studies. In 2002, Smidt et al. (2002a **R**) conducted another systematic review on the effectiveness of corticosteroid injections for lateral epicondylitis. This review found that corticosteroid injections had a positive short-term effect; however, due to the lack of high-quality studies, it was not possible to draw definitive conclusions.

In 1999, van der Windt et al. (1999 **R**) looked at the treatment effects of ultrasound therapy for musculoskeletal disorders. Thirty-eight studies were included in this review, but only six of these looked at lateral epicondylitis. The review concluded that there was little evidence to support the use of ultrasound therapy in the treatment of musculoskeletal disorders.

In 2002, Struijs et al. (2002 **R**) conducted a systematic review looking at the effects of orthotic devices for lateral epicondylitis. This study found that no definitive conclusions on orthotics could be made due to the methodological flaws present in the studies reviewed. A more recent review of orthotic devices was performed in 2004 (Borkholder et al. 2004 **R**). The authors conducted an exhaustive review of the literature, as well as a detailed analysis of the content and quality of available articles. For accurate comparison and consistency of terminology, splints described in the included articles were first classified according to the ASHT Splint Classification, and then according to their inherent material properties. Six splints in five classification categories were identified. Discussion of the results from the 11 studies that met minimum quality criteria was organised according to splint category and further separated into strength, pain, and load applied sections. This review identified one Sackett level 1b study and ten Sackett level 2b studies that offered early positive, but not conclusive, support for the effectiveness of splinting lateral epicondylitis. Limitations were noted in the way structure, fit, placement, and programmes of use were described, as well as in study quality. No specific type of orthotic was identified as being superior.

In 2001, Bernstein (2001 **R**) conducted a review to determine how effective surgical and injection therapy was in the management of chronic pain. This review found that local triamcinolone injection is effective for the relief of pain due to lateral epicondylitis (level 2). It was also found that there was limited evidence of effectiveness (level 3) for local glycosaminoglycan polyphosphate injection for lateral epicondylitis. Again, a lack of methodologically sound studies for surgery and injection therapies was noted.

In 2001, Mior (2001 **R**) looked at the effects of exercise in the treatment of chronic pain. This review only included one study looking at the upper extremity. This study found positive effects for exercise in the treatment of chronic lateral epicondylitis and for specific soft tissue shoulder disorders. However, due to the poor methodological quality of the study (level 3), definitive conclusions were not possible.

A review conducted in 2003 by Smidt et al. (2003 **R**) looked at the effectiveness of 'physiotherapy' for lateral epicondylitis. The study included twenty-three randomised controlled trials (RCTs) and found that two of the studies that compared ultrasound to a placebo ultrasound demonstrated statistically significant and clinically relevant differences in favor of ultrasound. There was, however, insufficient evidence to demonstrate either benefit or lack of effect for laser therapy, electrotherapy, exercises, and mobilisation techniques for lateral epicondylitis.

The most recent systematic review was conducted by ourselves (Trudel et al. 2004 **R**). A total of 209 studies were located, however, only 31 of these met the study inclusion criteria. Each of the articles was randomly allocated to reviewers and critically appraised using a structured critical appraisal tool with 23 items. Treatment recommendations were based on this rating and Sackett's level of evidence. We determined that level 2b evidence exists to support a number of treatments, including acupuncture, exercise therapy, manipulations and mobilisations, ultrasound, phonophoresis, Rebox, and ionisation with diclofenac. Each of these treatments had outcomes including either pain relief or improvement in function. There was also at least level 2b evidence showing laser therapy and pulsed electromagnetic field therapy are ineffective in the management of this condition.

Although this would suggest a promising slate of potential elements to a rehabilitation programme for our patient, there were noteworthy limitations in reviewed studies that indicated we should proceed with some scepticism. Many of the studies failed to provide adequate follow-up or blinding procedures, and used neither sample nor power calculations, nor sample size justification. The use of standardised outcome measures was another area of particular deficit. Recruitment strategies were often not described, making it difficult to generalise results; furthermore, the size and significance of effects were often absent. In addition, acute and chronic cases were rarely considered separately, either through stratification of sampling or in statistical analyses. A lack of clear descriptions of the techniques, dosages, and progressions, or of training and experience requirements, made it difficult for us to extract clear descriptions of the interventions used, even when reading the primary studies. Finally, despite the fact our patient has a work-relatedness problem, few studies address secondary prevention. The role for modification of workplace or recreation exposures was poorly studied. Therefore, when constructing the optimal treatment approach, we had to deal with uncertainty.

Despite this uncertainty, certain common elements appeared across the reviews. An active approach that includes exercise and education on self-management appears essential. For specific exercises, it may be necessary to delve into theoretical grounds and lower-quality studies as few studies are specific about the type, intensity, or duration of exercise. Rules for progressing exercise are rarely mentioned. However, we know that the size of the tendon is proportional to the size of the muscle, so muscle strengthening should increase tendon size and, hence, the ability of the tendon to resist the stresses of applied force. Progression of exercise should maximise tendon strength and functional endurance without increasing pain and potentially contributing to tendon pathology. Some modalities, such as ultrasound, have been shown to have positive effects (Binder et al. 1985 **A**; Lundeberg et al. 1988 **A**). It is noteworthy that some studies that have reported positive effects have applied ultrasound for 10 minutes (1 Mhz), so the specific parameters used should be matched to the original articles where possible. Acupuncture has shown positive short-term effects, but effects beyond 72 hours have not been identified. Our view on the use of modalities for pain relief is that they may be useful if they assist in achieving the core element of the programme (exercise, education, activity modification), but in isolation are not 'rehabilitation'. Given the pain scores reported by our patient and his use of anti-inflammatories, we believed we might be able to achieve our treatment objectives without a large emphasis on pain control. An orthosis might be useful, but no particular one has been identified as superior. Thus, either trial and error, theoretical or experiential approaches, or practical considerations would determine which device was selected. Outcome measures and patient feedback would be used to address the efficacy of specific orthotics for this particular patient. My particular rationale for selecting an orthotic is to choose a wrist cock-up splint where I feel that the tendons are inflamed or irritable and need rest, and to trial a counterforce type brace in more chronic or episodic cases as it may have an unloading effect without hampering function. This is an example of using level 5 evidence to make clinical decisions where clinical data is absent.

QUESTION 2

Which outcome measures used to monitor outcomes of tennis elbow rehabilitation might be useful for this patient?

We reviewed the outcomes used by clinical studies on tennis elbow and summarised our findings in table format to look for common measures and conceptual themes. The full table is available from the author. These data illustrated that even within clinical research studies, there is little consensus on the use of standardised outcome measures and a number of non-standardised measures continue to be used. Nevertheless, it was evident that some core constructs were being evaluated – pain, muscle strength, and function. We reviewed the literature (and used our clinical knowledge of the research on outcome measures) to try to identify an approach to core outcome measures for tennis elbow. We decided to differentiate short-term and long-term outcome constructs that were clinically relevant.

Based on this review, we proposed a strategy for evaluation of our patient, and of future patients with this condition, that includes the relevant concepts and viable options for measurement.

Outcomes

- Pain Relief (self-reported using either the Patient-Rated Tennis Elbow Evaluation (PRTEE) (MacDermid 2005 **C**; Newcomer et al. 2005 **A**; Overend et al. 1999 **A**), Pain-Free Function Questionnaire (Stratford et al. 1987 **A**), or a Visual Analogue Scale (VAS) or Numeric Pain Rating scale).
- Patient Function (using Patient-Rated Tennis Elbow Evaluation (PRTEE) (MacDermid 2005 **C**; Newcomer et al 2005**A**; Overend et al. 1999 **A**) or Disabilities of the Arm, Shoulder, Hand (DASH) (Beaton et al. 2001 **A**; Solway et al. 2002 **C**)).
- Muscle Function:
  - Functional grip – pain-free grip strength.
  - Tendon integrity – wrist extensor strength (depending on equipment availability).
  - Endurance for activity (a standardised test has yet to be described, so not a viable option at this time).

Long-term outcomes

- Reoccurrence of Symptoms:
  - Pain/function (using Patient-Rated Forearm Scale (MacDermid 2005 **C**; Newcomer et al. 2005 **A**; Overend et al.1999 **A**).
  - Requirement of additional treatment.
- Work Outcomes (measured by lost time, the Work subscale of the DASH (Beaton et al. 2001 **A**; Solway et al. 2002 **C**), or a scale similar to the Work Limitation Questionnaire (WLQ) (Lerner et al. 2001 **A**; Lerner et al. 2002 **A**), which describes difficulty at work).
- Resumption of Valued Regular Recreational Activity.

Self-report scales designed specifically for patients with lateral epicondylitis are available and are likely to be most responsive to changes in LE symptoms (Newcomer et al. 2005 **A**; Stratford et al. 1993 **A**). The Patient-Rated Tennis Elbow Evaluation (PRTEE) has pain and function (specific and usual activity) subscales, which are weighted equally to provide a global score (range 0–100; 100 worst) (MacDermid 2005 **C**). The Pain-Free Function Questionnaire is a pain scale that focuses on pain with activity (Stratford et al. 1987 **A**). Both were developed with items specific to lateral epicondylitis. Other self-report measures with sound psychometric properties, such as the Disabilities of the Arm, Shoulder and Hand (DASH), the Numeric Rating Scale for pain (Ferraz et al. 1990 **A**; Jaeschke et al. 1990 **A**; Scudds 2001 **R**), or the McGill Pain Questionnaire (Melzack 1975 **A**, 1987 **A**) might also contribute to a more comprehensive comparison of treatment interventions, but are less specific to the condition. However, as head-to-head evaluations of these different outcome measures have not yet been performed, their relative measurement properties are unknown. In terms of measuring physical impairments, both ROM and strength measures have been studied (Pienimaki et al. 2002 **A**; Smidt et al. 2002b **A**; Stratford et al. 1993 **A**). Pain-free grip (measured with the elbow extended) has been shown to be reliable, valid, and responsive in this patient population (Smidt et al. 2002b **A**; Stratford et al. 1987 **A**; Wuori et al.1998 **A**; Overend et al. 1998). Pain threshold can be measured by algometry, although this may be less reliable than other physical measures (Smidt et al. 2002b **A**). Based on our case and the importance of work outcomes, we chose pain-free grip, the PRTEE, and the WLQ as outcomes to monitor the impact of our programme.

QUESTION 3

What is the optimal method for assessing strength with this problem?

We found that pain-free grip was commonly used in outcome studies, and there were studies suggesting it is better than other indicators, notably maximum grip strength, in detecting change over time (Stratford et al. 1993 **A**). The intraclass correlation coefficients (ICCs) for the pain-free grip strength and maximum grip strength were 0.97 and 0.98 respectively, indicating excellent reliability (Smidt et al. 2002b **R**; Stratford et al. 1989 **A**) in this patient population. Pain-free grip measurement uses a different methodology to that recommended by the ASHT (Fess 1992 **C**) for maximum grip strength testing, and the following variations are to be incorporated: 1. the elbow is fully extended (not at 90 degrees), and 2. the patient is asked to grip as hard as they can without causing pain. In my own (level 5) clinical experience, I have found comparing the maximum and pain-free grip strength to be informative, although little research has specifically addressed whether the gap between maximum and pain-free grip strength is a useful measure of tissue irritability. As I find no literature supporting or refuting that premise, I remain sceptical, but am not yet prepared to reject the comparison.

QUESTION 4

What factors modify the prognosis for recovery and return to work following tennis elbow?

Searching for 'prognosis', 'rehabilitation', and 'lateral epicondylitis' (or tennis elbow), we identified two relevant studies. A systematic review conducted by Hudak et al. (1996 **R**) was unable to reach clear conclusions because estimates of duration were only available from weaker studies with longer follow-up times; significant subject heterogeneity existed and this prevented a determination of a usual clinical course. There was limited evidence that site of pathology might influence outcomes (Hudak et al. 1996 **R**). A more recent study evaluated prognosis in 83 patients attending an eight week physiotherapy programme for management of unilateral lateral epicondylitis. The final prognostic model for pain and disability, measured using the DASH scores, included the baseline DASH score, sex (female), and self-reported nerve symptoms. A sub-analysis indicated that women were more likely than men to have work-related onsets, repetitive keyboarding jobs, and cervical joint signs. Among women, these factors were associated with higher final DASH and VAS scores. While the data were not all directly applicable to our male patient, this study suggested we should examine for nerve symptoms and consider work issues. His initial DASH score of 30 was favourable as it was about 20 points less than in patients with other upper extremity disorders who were unable to return to work (Beaton et al. 2001 **A**). While not specific to lateral epicondylitis, early intervention that addresses both physical and psychosocial stressors at work has been suggested as necessary in UED (Feuerstein et al. 2000 **A**; Feuerstein et al. 2004 **A**; Himmelstein et al. 1995 **A**; Huang et al. 2002 **A**; Shaw et al. 2001 **A**).

CASE REPORT II

Mrs CT is a 56 year old, right-handed female who works as an accounting clerk. She self-referred to the clinic. During the initial interview, she stated that she has tingling in her fingers (right hand) that is worse at night and has been present for three months. She wakes two to three times each night with this problem, which resolves when she shakes her hand. This is very similar to her experience when she was pregnant (20 years ago). That episode receded with the birth of her baby. She has some achy pain that is hard to localise, and the days when she is required to sort through files seem to make things worse. Other than this problem, she is in good health.

PERTINENT FINDINGS ON PHYSICAL EXAMINATION

- Joint Motion: wrist and hand within normal limits.
- Static Muscle Testing: resisted thumb abduction – pain free; normal power.
- Measured Strength:
  - Left: maximum grip 34 kg; tripod pinch 4 kg.
  - Right: maximum grip 23 kg; tripod pinch 4.5 kg.
- Sensation: Semmes-Weinstein Monofilament testing (SWMF) R D3 = 3.22; L D3 = 2.83; R D5 = 2.83.
- Special Tests:
  - Wrist flexion test: positive in 15 seconds on right; negative on left.
  - Tinel's test: positive on right; negative on left.
  - Allen's test: negative both sides.
  - Cervical compression test: negative.
  - Cervical quadrant tests: negative both sides.

Carpal tunnel syndrome (CTS) has the highest prevalence of all forms of compression neuropathy (Atroshi et al. 1999 **A**; Stevens et al. 1988 **A**; Zakaria 2004 **A**). The median nerve is susceptible to pressure as it passes, with the flexor tendons, through the carpal tunnel in a space defined by the concave arch of the carpus and enclosed by the transverse carpal ligament (TCL) (Mesgarzadeh et al. 1989 **A**; Rotman & Donovan 2002 **C**). The palmar wrist crease corresponds to the proximal border of the TCL and the TCL attaches medially to the pisiform and hamate and laterally to the scaphoid tuberosity and trapezium. The median nerve normally enters the carpal tunnel in the midline or slightly radial to it. The thenar branch most commonly separates from the median nerve distal to the transverse carpal ligament, but can branch off within the carpal tunnel. Sensory branches supply the radial 3 $1/2$ digits. However, the cutaneous skin of the palm is supplied by the palmar sensory cutaneous branch of the median nerve, which arises, on average, 6 cm proximal to the TCL and, therefore, should not be affected in CTS.

Nerve fibres have layers of connective tissue. The extensibility of these layers is critical to nerve gliding. It has been demonstrated that the median nerve will move up to 9.6 mm with flexion and slightly less with wrist extension (Tuzuner et al. 2004 **A**; Wright et al. 1996 **A**). Chronic compression is thought to cause fibrosis, which will inhibit nerve gliding. Injury/scarring of the mesoneurium will cause the nerve to adhere to surrounding tissue. This may result in traction of the nerve during movement, as the nerve attempts to glide from this fixed position.

The pathophysiology of nerve compression, and how it relates to evaluation and treatment, has been well described by MacKinnon (2002 **B**). The pathophysiology of Grade 1 nerve injury (neuropraxia) involves conduction block and may be associated with some segmental areas of demyelination. The axon is not injured and does not undergo regeneration. A Grade 2 nerve injury (axonotmesis) involves injury to the axon itself. The nerve will have lost some fibres and be in a process of nerve repair. Despite these changes, this injury also has potential to recover completely. A Grade 3 injury has both loss of axons and some degree of scar tissue in the endoneurium. Patients with such an injury will have constant numbness and observable thenar atrophy. These patients have severe carpal tunnel syndrome and complete recovery may not be achievable. Grades 4 and 5 involve complete scarring or transaction of the nerve and do not apply to CTS. Understanding the factors that contribute to increased pressure in the carpal tunnel, including the anatomy, posture, size of enclosed structures (tendon, nerve), and vascular components of pressure, is fundamental to defining treatment programmes. The severity of the compression determines which diagnostic tests are most likely to be positive, which treatments will be effective, and relates to overall prognosis.

QUESTION 1

What clinical tests are useful for diagnosis of carpal tunnel syndrome?

Two systematic reviews have been conducted on clinical diagnostic tests for CTS (MacDermid & Wessel 2004 **R**; Massy-Westropp et al. 2000 **R**). Our study was more recent, exceeded the 21 papers reported upon in a previous systematic review (Massy-Westropp et al. 2000 **R**), and used rigorous search and appraisal methods. Thus, we relied on this review to provide an overview of the numerous studies that present conflicting results on test validity. The controversy over the value of clinical tests was not surprising when we realised that the majority of studies failed to report on the diagnosis of the subjects without CTS or on the reliability of the diagnostic tests (MacDermid & Wessel 2004 **R**). Furthermore, few studies indicated whether the testers were blinded to the gold standard results, and less than half described non-cases representing the spectrum of patients who would normally present for differential diagnosis. Only 15 studies (Atroshi et al. **A**; Breidenbach & McCabe 1997 **A**; Bland 2000 **A**; Cherniack et al. 1996 **A**; Fertl et al.1998 **A**; Ghavanini & Haghighat 1998 **A**; Gunnarsson et al. 1997 **A**; Karl et al. 2001 **A**; Kaul et al. 2000 **A**; Kaul et al. 2001 **A**; Kuhlman & Hennessey 1997 **A**; MacDermid et al. 1994 **A**; MacDermid et al. 1997 **A**; Pagel et al. 2002 **A**; Pryse-Phillips 1984 **A**; Walters & Rice 2002 **A**) had quality scores indicating that eight or more key quality indicators were met (out of 12).

Given the variety of diagnostic tests for CTS, the large number of studies, and the widely disparate results, it was difficult to make firm conclusions on the value of specific tests. Therefore, we classified tests into three groups: 'Unable to Make Recommendations', 'Not Useful', and 'Potentially Useful', in our review of clinical diagnosis of CTS. Potentially useful tests included Phalen's, Tinel's, Carpal Compression (CC), Wrist Extension, CC + Wrist Flexion, Flick Sign, Gilliat Tethered Median Nerve, Hand Diagram, Fist (Lumbrical Provocation), Static 2-point, Abductor Pollicus Brevis (APB) Strength, APB atrophy, Current Perception threshold, Semmes-Weinstein monofilament, Vibration Threshold Testing (with tuning fork or vibrometer). We also used a simplistic 'meta-analysis' strategy, in which we combined estimated sensitivity and specificity reported for individual tests across studies weighted by sample size. While our meta-analysis strategy was simplistic, we felt it was necessary to provide more stable estimates, particularly on test sensitivity. For example, it is difficult to make decisions based on the numerous studies evaluating Phalen's (wrist flexion) test, given that sensitivity ranges from 10 % to 91 % (MacDermid 1991 **R**). The sensitivity of 68 % achieved over 3,000 cases provides relatively strong evidence that this test is useful, although false negatives can be anticipated. A previous review (Massy-Westropp et al. 2000 **R**) suggested that 2-point discrimination is specific but not sensitive; we were able to confirm these characteristics. Across six studies and over 500 patients, specificity was 95 %, while sensitivity was only 24 %.

Given the number of tests, we refer the reader to our tables in the systematic review (MacDermid & Wessel 2004 **R**) and subsequent narrative review (MacDermid & Doherty 2004 **R**), where we describe our results in detail. We were able to sort out which tests were more sensitive and which more specific, allowing us to make recommendations on how particular test results might be interpreted. We devised a summary of which test results (+ or −) provide strong (++) or weak (+) evidence to support a diagnosis of CTS. We also indicated test results that have no effect on the expected correct clinical diagnosis (0), and others that reduce the probability of CTS being the correct diagnosis (see Table 9.1). This example illustrates where a

Table 9.1. Steps towards a conclusive clinical diagnosis: the influence of different test outcomes on likelihood of CTS.

Define the Nature of the Symptoms				
++	+	0	−	−−
Paresthesia, numbness and pain. Focal swelling just proximal to wrist crease. Waking at night with Paresthesia.	Hand swelling. Symptoms relieved by flicking of wrists. Paresthesia with activity or position.	Pain aggravated by movement or position.	Pain only.	

Define Location of Sensory Complaints				
++	+	0	−	−−
D1–D3 included. Ring-splitting. Exclusion of D5. Exclusion of palm.	Symptoms in 1 or more radial digits.	Diffuse including hand.	D5 involved. Include palm (implicates forearm). Radiate proximal to wrist.	Symptoms follow dermatome (implicates neck). Extend into forearm (implicates forearm). D5 only (implicates ulnar nerve).

Sensory Examination				
++	+	0	−	−−
Abnormal threshold (vibration, SWMF, current perception) in D1–D3 with Normal D5. Abnormal 2-point D1–D3.	Abnormal threshold in at least 1 of D1–D3.	Normal 2-point in digits.	Abnormal threshold D5 (ulnar nerve +).	Normal threshold in D1–D3.

Motor Examination				
++	+	0	−	−−
Weak abduction of thumb. Atrophy of thenar bulk.		Decreased grip strength, grip endurance. Normal thenar bulk.	Proximal/thenar weakness (+ forearm, neck or disuse atrophy).	Proximal atrophy (neck/brachial plexus). Abnormal reflexes (neck).

Table 9.1. (*Continued*)

Special Tests				
++	+	0	−	−−
Wrist flexion + Carpal compression+ Nerve percussion +	Percussion−		Wrist flexion or Carpal Compression−	Carpal compression and wrist flexion−

Response to Night Splints				
++	+	0	−	−−
Reduced symptoms.				No reduction in symptoms.

systematic review (MacDermid & Wessel 2004 **R**) and a narrative review based on a systematic review (MacDermid & Doherty 2004 **R**) can be helpful in reaching useful conclusions where evidence is overwhelming because of its depth and lack of clarity.

QUESTION 2

What is the best approach for successful conservative management in carpal tunnel syndrome?

We used a systematic review to devise an overall approach to CTS. The best available evidence to date shows significant benefits (Grade B recommendations) from splinting, ultrasound, nerve gliding exercises, carpal bone mobilisation, magnetic therapy, and yoga for people with CTS. The evidence also indicates that the effects of ultrasound or magnetic therapy depend on specific treatment parameters; pulsed, deep ultrasound or prolonged magnetic therapy is effective, while continuous, superficial ultrasound or brief single-session magnetic therapy is not. There is some evidence (Grade C recommendation) to support the use of laser therapy and various combined therapies. However, results from acupuncture research are inconclusive (Grade D). The detailed summary of these studies is published elsewhere (Muller et al. 2004 **R**), so the reader can compare studies to assess similarity to our patient.

The results of our systematic review (Muller et al. 2004 **R**) suggest that there are many conservative physiotherapy interventions that could be used in the treatment of CTS. An earlier review also concluded that there is evidence for significant short-term benefit from oral steroids, splinting, ultrasound, yoga (a type that emphasises movement), and carpal bone mobilisation (O'Connor et al. 2003 **A**), although another review suggested that steroid injection may have a larger impact (Gerritsen et al. 2001 **R**). Evidence-based practice combines the results of research trials with the unique presentation and needs of the individual patient. Choosing interventions that have proven effective in subjects who present similarly to the patient in question will likely improve the potential for a positive outcome. Our patient was a female with a previous history of CTS that responded to splinting. Given the support for splinting in the literature, we would proceed with night splinting (wrist in neutral) and provide gliding exercises and education on activity modification. As improvement in symptoms is expected within three to six weeks, we would review whether other interventions were necessary at three weeks. At this time carpal bone mobilisation might be added to the programme – if the therapist had the required skill.

In reviewing the literature we noted that a novel splint was reported to be effective (Manente et al. 2001 **A**), but felt that evidence was too preliminary to proceed with that option as a first line choice. We decided that if our splinting programme was not as successful, we might consider it as a second attempt. Given our lack of certainty, we could use an N of 1 trial design (Cook 1996 **A**; Mahon et al. 1996 **A**; Rodnick 2006 **C**) to evaluate the use of this splint for this particular patient. We might have used a similar approach in our previous problem to determine which orthotic was best suited to our tennis elbow patient. N of 1 trials offer a rigorous method for dealing with uncertainty in individual patients, as different treatment components can be evaluated in terms of their effectiveness for a single patient.

QUESTION 3

Which self-report outcome measure would be most useful for detecting change in carpal tunnel symptoms following treatment?

We decided that because CTS is a syndrome characterised by specific symptoms, a change in these symptoms would be a useful clinical indicator of success. While we found in our search a variety of functional scales, we were attracted to the Symptom Severity Scale (SSS) described by Levine et al. (1993 **A**) as it clearly focused on the primary symptoms our patient was experiencing. Our concern was – *Is this measure reliable and valid, particularly in comparison to other potential scales that emphasise hand function?* We searched the literature for mentions of the scale (noting the various names that are used in the literature, including Symptom Severity Scale, Boston Carpal Tunnel Scale, Brigham and Women's, and Levine's). We found a number of articles that address reliability, validity, and responsiveness, and all agree that the SSS is at least as responsive, if not more responsive than comparative measures, and that it has high reliability (Amadio et al. 1996 **A**; Atroshi et al. 1998 **A**; Bessette et al. 1998 **A**; Gay et al. 2003 **A**).

In our review of the literature, we also found a table describing scores for patients who proceeded to surgery following conservative management as compared to those who did not, and retrieved this information. We compared scores reported in other

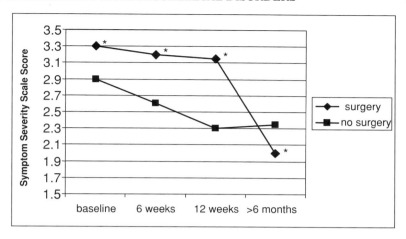

Figure 9.1. CSSS reported by a cohort of patients treated with 12 weeks of night-splinting, comparing the scores of those who were successful with conservative management with those who proceeded to have a carpal tunnel release (*) (>6-month post-op scores for surgical group). Statistically significant (p < 0.05) difference between the group that had surgery and the group that did not at every time point. Data adapted from that reported by others (Boyd et al. 2005 **A**).

studies with our patient's score at baseline and follow-up, to assess severity and response to treatment (see Figure 9.1).

QUESTION 4

Does a positive Tinel's score on the Symptom Severity Scale have prognostic value?

Based on our knowledge of the pathophysiology of nerve compression (Mackinnon 2002 **B**), we had reason to believe that a positive Tinel's test might indicate more severe CTS, for example, axonoteomesis. If this was true then response to splinting might be slower or less successful. In searching the literature, we found that it had been confirmed in clinical studies that Tinel's was more likely to be positive in later stages of compression (Novak et al. 1992 **A**). We also found a study that evaluated prognosis to three weeks of splinting when combined with steroid injection (Gelberman et al. 1980 **A**). This study reported that patients that initially had mild symptoms of less than one year's duration, normal sensibility, normal thenar strength and mass, and one or two millisecond prolongations of either distal median motor or sensory latencies, had the most satisfactory responses to injections and splinting. Patients with severe symptoms of more than one year's duration, findings of atrophy and weakness, and distal motor latencies of more than six milliseconds or absent sensory responses, had the poorest response to injections and experienced a high rate of relapse (Gelberman et al. 1980 **A**). A similar study conducted on 50 'hands' in 34 patients, followed patients for 18 months after steroid injection with splinting (Stahl et al. 1996 **A**). Conservative therapy was effective in 82 % of hands after eight weeks, but symptoms tended to recur so that by the end of a year only 20 % remained asymptomatic. Failure of conservative therapy was predicted by long duration of symptoms, older age, permanent paresthesia, 2-point discrimination threshold above 6 mm, positive Phalen's test within 30 seconds, and long motor and sensory distal latency. Other studies have shown that some benefit in reduced symptoms exists for workers, even if median nerve changes are evident – although relief will not be complete (Werner et al. 2005 **A**). Finally, our clinicians conducted a study looking at the impact of SSS on likelihood to proceed to surgery following conservative management, and found that patients who proceeded to surgery were characterised by having higher SSS and a failure to improve within the first six weeks (Boyd et al. 2005 **A**) (see Figure 9.1). Our patient's score of 3.0 was consistent with response to conservative management. We concluded that our patient has some risk of failure to respond to our initial treatment programme. Risk factors included a positive Tinel's, a positive wrist flexion test in less than 30 seconds, recurrence of symptoms, and a moderate to high SSS. We decided to conduct a more detailed job analysis to mitigate risk as much as possible, and to follow her at both three and six weeks to re-evaluate response to treatment. We informed her that if the splint did not completely resolve her symptoms, it was still likely to improve them; however, it might be necessary to try other treatments and to re-evaluate the need for surgery in the future.

CONCLUSION

These examples do not provide a comprehensive view of UED. They do show the approach to delivering evidence-based management of two common UEDs. As the therapist continues to use this approach across different cases and conditions, principles emerge, and clarity on the ideal approach for many UEDs will crystallise. This is the difference between '20 years of practice and one year of practice repeated 20 times'. An ongoing process of using the best and latest knowledge to support the treatment choices made, and an associated valid process for evaluating the impact of those choices, will provide a foundation for enhanced expertise in managing UED.

REFERENCES

Aaras A, Horgen G, Bjorset HH, Ro O, Walsoe H (2001) Musculoskeletal, visual and psychosocial stress in VDU operators before and after multidisciplinary ergonomic interventions. A 6 years prospective study – Part II. *Applied Ergonomics* **32**(6): 559–571.
Amadio PC, Silverstein MD, Ilstrup DM, Schleck CD, Jensen LM (1996) Outcome assessment for carpal tunnel surgery: the relative responsiveness of generic, arthritis-specific, disease-specific, and physical examination measures. *Journal of Hand Surgery, American Volume* **21**(3): 338–346.

Assendelft WJ, Hay EM, Adshead R, Bouter LM (1996) Corticosteroid injections for lateral epicondylitis: a systematic overview. *British Journal of General Practice* **46**(405): 209–216.

Atroshi I, Breidenbach WC, McCabe SJ (1997) Assessment of the carpal tunnel outcome instrument in patients with nerve-compression symptoms. *Journal of Hand Surgery, American Volume* **22**(2): 222–227.

Atroshi I, Gummesson C, Johnsson R, Ornstein E, Ranstam J, Rosen I (1999) Prevalence of carpal tunnel syndrome in a general population. *Journal of the American Medical Association* **282**(2): 153–158.

Atroshi I, Johnsson R, Sprinchorn A (1998) Self-administered outcome instrument in carpal tunnel syndrome. Reliability, validity and responsiveness evaluated in 102 patients. *Acta Orthopaedica Scandinavica* **69**(1): 82–88.

Baker NA, Jacobs K, Carifio J (1999) The ability of background factors, work practices, and psychosocial variables to predict the severity of musculoskeletal discomfort. *Occupational Ergonomics* **2**(1): 27–41.

Beaton DE, Katz JN, Fossel AH, Wright JG, Tarasuk V, Bombardier C (2001) Measuring the whole or the parts? Validity, reliability, and responsiveness of the Disabilities of the Arm, Shoulder and Hand outcome measure in different regions of the upper extremity *Journal of Hand Therapy* **14**(2): 128–146.

Bernstein RM (2001) Injections and surgical therapy in chronic pain *Clinical Journal of Pain* **17**(4); Suppl: 94S–104S.

Bessette L, Sangha O, Kuntz KM, Keller RB, Lew RA, Fossel AH et al. (1998) Comparative responsiveness of generic versus disease-specific and weighted versus unweighted versus un-weighted health status measures in carpal tunnel syndrome. *Medical Care* **36**(4): 491–502.

Binder A, Hodge G, Greenwood AM, Hazleman BL, Page Thomas DP (1985) Is therapeutic ultrasound effective in treating soft tissue lesions? *British Medical Journal (Clinical Research Edition)* **290**(6467): 512–514.

Bland JD (2000) The value of the history in the diagnosis of carpal tunnel syndrome. *Journal of Hand Surgery, British & European Volume* **25**(5): 445–450.

Bongers PM, Kremer AM, ter Laak J (2002) Are psychosocial factors, risk factors for symptoms and signs of the shoulder, elbow, or hand/wrist?: a review of the epidemiological literature. *American Journal of Industrial Medicine* **41**(5): 315–342.

Borkholder CD, Hill VA, Fess EE (2004) The efficacy of splinting for lateral epicondylitis: a systematic review. *Journal of Hand Therapy* **17**(2): 181–199.

Boyd KU, Gan BS, Ross DC, Richards RS, Roth JH, MacDermid JC (2005) Outcomes in carpal tunnel syndrome: symptom severity, conservative management and progression to surgery. *Clinical and Investigative Medicine* **28**(5): 254–260.

Cherniack MG, Moalli D, Viscolli C (1996) A comparison of traditional electrodiagnostic studies, electroneurometry, and vibrometry in the diagnosis of carpal tunnel syndrome. *Journal of Hand Surgery, American Volume* **21**(1): 122–131.

Cook DJ (1996) Randomized trials in single subjects: the N of 1 study. *Psychopharmacology Bulletin* **32**(3): 363–367.

de Jonge J, Reuvers MM, Houtman IL, Bongers PM, Kompier MA (2000) Linear and nonlinear relations between psychosocial job characteristics, subjective outcomes, and sickness absence: baseline results from SMASH. Study on Musculoskeletal Disorders, Absenteeism, Stress, and Health. *Journal of Occupational Health Psychology* **5**(2): 256–268.

Devereux JJ, Vlachonikolis IG, Buckle PW (2002) Epidemiological study to investigate potential interaction between physical and psychosocial factors at work that may increase the risk of symptoms of musculoskeletal disorder of the neck and upper limb. *Journal of Occupational and Environmental Medicine* **59**(4): 269–277.

Ferraz MB, Quaresma MR, Aquino LR, Atra E, Tugwell P, Goldsmith CH (1990) Reliability of pain scales in the assessment of literate and illiterate patients with rheumatoid arthritis. *Journal of Rheumatology* **17**(8): 1022–1024.

Fertl E, Wober C, Zeitlhofer J (1998) The serial use of two provocative tests in the clinical diagnosis of carpal tunnel syndrome. *Acta Neurologica Scandinavica* **98**(5): 328–332.

Fess EE (1992) Grip strength. In: Casanova JS (ed.) *Clinical Assessment Recommendations* (2 edn) Chicago: American Society of Hand Therapists, pp. 41–46.

Feuerstein M, Huang GD, Haufler AJ, Miller JK (2000) Development of a screen for predicting clinical outcomes in patients with work-related upper extremity disorders. *Journal of Occupational and Environmental Medicine* **42**(7): 749–761.

Feuerstein M, Shaw WS, Nicholas RA, Huang GD (2004) From confounders to suspected risk factors: psychosocial factors and work-related upper extremity disorders. *Journal of Electromyography and Kinesiology* **14**(1): 171–178.

Gay RE, Amadio PC, Johnson JC (2003) Comparative responsiveness of the disabilities of the arm, shoulder, and hand, the carpal tunnel questionnaire, and the SF-36 to clinical change after carpal tunnel release. *Journal of Hand Surgery, American Volume* **28**(2): 250–254.

Gelberman RH, Aronson D, Weisman MH (1980) Carpal-tunnel syndrome. Results of a prospective trial of steroid injection and splinting. *Journal of Bone and Joint Surgery, American Volume* **62**(7): 1181–1184.

Gerritsen AA, Uitdehaag BM, van Geldere D, Scholten RJ, de Vet HC, Bouter LM (2001) Systematic review of randomized clinical trials of surgical treatment for carpal tunnel syndrome. *British Journal of Surgery* **88**(10): 1285–1295.

Ghavanini MR, Haghighat M (1998) Carpal tunnel syndrome: reappraisal of five clinical tests. *Electromyography and Clinical Neurophysiology* **38**(7): 437–441.

Gunnarsson LG, Amilon A, Hellstrand P, Leissner P, Philipson L (1997) The diagnosis of carpal tunnel syndrome. Sensitivity and specificity of some clinical and electrophysiological tests. *Journal of Hand Surgery, British & European Volume* **22**(1): 34–37.

Himmelstein JS, Feuerstein M, Stanek EJ, Koyamatsu K, Pransky GS, Morgan W et al. (1995) Work-related upper-extremity disorders and work disability: clinical and psychosocial presentation. *Journal of Occupational and Environmental Medicine* **37**(11): 1278–1286.

Huang GD, Feuerstein M, Kop WJ, Schor K, Arroyo F (2003) Individual and combined impacts of biomechanical and work organization factors in work-related musculoskeletal symptoms. *American Journal of Industrial Medicine* **43**(5): 495–506.

Huang GD, Feuerstein M, Sauter SL (2002) Occupational stress and work-related upper extremity disorders: concepts and models. *American Journal of Industrial Medicine* **41**(5): 298–314.

Hudak PL, Cole DC, Haines AT (1996) Understanding prognosis to improve rehabilitation: the example of lateral elbow pain. *Archives of Physical Medicine and Rehabilitation* **77**(6): 586–593.

Jaeschke R, Singer J, Guyatt GH (1990) A comparison of seven-point and visual analogue scales: data from a randomized trial. *Controlled Clinical Trials* **11**: 43–51.

Johansson JA, Rubenowitz S (1994) Risk indicators in the psychosocial and physical work environment for work-related neck, shoulder and low back symptoms: a study among blue- and white-collar workers in eight companies. *Scandinavian Journal of Rehabilitation Medicine* **26**(3): 131–142.

Karl AI, Carney ML, Kaul MP (2001) The lumbrical provocation test in subjects with median inclusive paresthesia. *Archives of Physical Medicine and Rehabilitation* **82**(7): 935–937.

Kaul MP, Pagel KJ, Dryden JD (2000) Lack of predictive power of the tethered median stress test in suspected carpal tunnel syndrome. *Archives of Physical Medicine and Rehabilitation* **81**(3): 348–350.

Kaul MP, Pagel KJ, Wheatley MJ, Dryden JD (2001) Carpal compression test and pressure provocative test in veterans with median-distribution paresthesias. *Muscle Nerve* **24**(1): 107–111.

Kraushaar BS, Nirschl RP (1999) Current concepts review: tendinosis of the elbow (tennis elbow). Clinical features and findings of histological, immunohistochemical, and electron microscopy studies. *Journal of Bone and Joint Surgery, American Volume* **81**(2): 259–278.

Kuhlman KA, Hennessey WJ (1997) Sensitivity and specificity of carpal tunnel syndrome signs. *American Journal of Physical Medicine and Rehabilitation* **76**(6): 451–457.

Lerner D, Amick BC III, Rogers WH, Malspeis S, Bungay K, Cynn D (2001) The work limitations questionnaire. *Medical Care* **39**(1): 72–85.

Lerner D, Reed JI, Massarotti E, Wester LM, Burke TA (2002) The work limitations questionnaire's validity and reliability among patients with osteoarthritis. *Journal of Clinical Epidemiology* **55**(2): 197–208.

Levine DW, Simmons SP, Koris MJ, Daltroy LH, Hohl GG, Fossel AH et al. (1993) A self-administered questionnaire for assessment of severity of symptoms and functional status in carpal tunnel syndrome. *Journal of Bone and Joint Surgery, American Volume* **75A**(11): 1585–1592.

Lundberg U (2002) Psychophysiology of work: stress, gender, endocrine response, and work-related upper extremity disorders. *American Journal of Industrial Medicine* **41**(5): 383–392.

Lundeberg T, Abrahamsson P, Haker E (1988) A comparative study of continuous ultrasound, placebo ultrasound and rest in epicondylalgia. *Scandinavian Journal of Rehabilitation Medicine* **20**(3): 99–101.

MacDermid J (2005) Update: the patient-rated forearm evaluation questionnaire is now the patient-rated tennis elbow evaluation. *Journal of Hand Therapy* **18**(4): 407–410.

MacDermid JC (1991) Accuracy of clinical tests used in the detection of carpal tunnel syndrome: a literature review. *Journal of Hand Therapy* **4**: 169–176.

MacDermid JC (2004) The quality of clinical practice guidelines in hand therapy. *Journal of Hand Therapy* **17**(2): 200–209.

MacDermid JC, Doherty T (2004) Clinical and electrodiagnostic testing of carpal tunnel syndrome: a narrative review. *Journal of Orthopaedic and Sports Physical Therapy* **34**(10): 565–588.

MacDermid JC, Kramer JF, McFarlane RM, Roth JH (1997) Inter-rater agreement and accuracy of clinical tests used in diagnosis of Carpal Tunnel Syndrome. *WORK: a journal of prevention, assessment & rehabilitation* **8**(1): 37–44.

MacDermid JC, Kramer JF, Roth JH (1994) Decision making in detecting abnormal Semmes-Weinstein monofilament thresholds in carpal tunnel syndrome. *Journal of Hand Therapy* **7**(3): 158–162.

MacDermid JC, Wessel J (2004) Clinical diagnosis of carpal tunnel syndrome: a systematic review. *Journal of Hand Therapy* **17**(2): 309–319.

Mackinnon SE (2002) Pathophysiology of nerve compression. *Hand Clinics* **18**(2): 231–241.

Mahon J, Laupacis A, Donner A, Wood T (1996) Randomised study of n of 1 trials versus standard practice. *British Medical Journal* **312**(7038): 1069–1074.

Manente G, Torrieri F, Di Blasio F, Staniscia T, Romano F, Uncini A (2001) An innovative hand brace for carpal tunnel syndrome: a randomized controlled trial. *Muscle Nerve* **24**(8): 1020–1025.

Massy-Westropp N, Grimmer K, Bain G (2000) A systematic review of the clinical diagnostic tests for carpal tunnel syndrome. *Journal of Hand Surgery, American Volume* **25**(1): 120–127.

Melzack R (1975) The McGill pain questionnaire: major properties and scoring methods. *Pain*: 277–299.

Melzack R (1987) The short-form McGill pain questionnaire. *Pain* **30**: 191–197.

Mesgarzadeh M, Schneck CD, Bonakdarpour A (1989) Carpal tunnel: MR imaging. Part I. Normal anatomy. *Radiology* **171**: 743–748.

Mior S (2001) Exercise in the treatment of chronic pain. *Clinical Journal of Pain* **17**(4); Suppl: 77S–85S.

Muller M, Tsui D, Schnurr R, Biddulph-Deisroth L, Hard J, MacDermid JC (2004) Effectiveness of hand therapy interventions in primary management of carpal tunnel syndrome: a systematic review. *Journal of Hand Therapy* **17**(2): 210–228.

Newcomer KL, Martinez-Silvestrini JA, Schaefer MP, Gay RE, Arendt KW (2005) Sensitivity of the Patient-Rated Forearm Evaluation Questionnaire in lateral epicondylitis. *Journal of Hand Therapy* **18**(4): 400–406.

Novak CB, Mackinnon SE (2002) Multilevel nerve compression and muscle imbalance in work-related neuromuscular disorders. *American Journal of Industrial Medicine* **41**(5): 343–352.

Novak CB, Mackinnon SE, Brownlee R, Kelly L (1992) Provocative sensory testing in carpal tunnel syndrome. *Journal of Hand Surgery, British & European Volume* **17**(2): 204–208.

O'Connor D, Marshall S, Massy-Westropp N (2003) Non-surgical treatment (other than steroid injection) for carpal tunnel syndrome. *Cochrane Library* **1** http://www.thecochranelibrary.com CD003219.

Overend TJ, Wuori-Fearn JL, Kramer JF, MacDermid JC (1999) Reliability of a patient-rated forearm evaluation questionnaire for patients with lateral epicondylitis. *Journal of Hand Therapy* **12**(1): 31–37.

Pagel KJ, Kaul MP, Dryden JD (2002) Lack of utility of Semmes-Weinstein monofilament testing in suspected carpal tunnel syndrome. *American Journal of Physical Medicine and Rehabilitation* **81**(8): 597–600.

Pienimaki TT, Siira PT, Vanharanta H (2002) Chronic medial and lateral epicondylitis: a comparison of pain, disability, and function. *Archives of Physical Medicine and Rehabilitation* **83**(3): 317–321.

Pryse-Phillips WE (1984) Validation of a diagnostic sign in carpal tunnel syndrome. *Journal of Neurology, Neurosurgery and Psychiatry* **47**(8): 870–872.

Rodnick JE (2006) Australia: the N of 1 trial, an underappreciated research method. *Family Medicine* **38**(1): 63.

Rotman MB, Donovan JP (2002) Practical anatomy of the carpal tunnel. *Hand Clinics* **18**(2): 219–230.

Sackett DL, Straus SE, Richardson WS, Rosenberg W, Haynes RB (2000) *Evidence-Based Medicine: how to practice and teach EBM* (2 edn) Toronto, Canada: Churchill Livingstone.

Scudds RA (2001) Pain outcome measures. *Journal of Hand Therapy* **14**(2): 86–90.

Shaw WS, Feuerstein M, Miller VI, Lincoln AE (2001) Clinical tools to facilitate workplace accommodation after treatment for an upper extremity disorder. *Assistive Technology* **13**(2): 94–105.

Smidt N, Assendelft WJ, Arola H, Malmivaara A, Greens S, Buchbinder R et al. (2003) Effectiveness of physiotherapy for lateral epicondylitis: a systematic review. *Annals of Medicine* **35**(1): 51–62.

Smidt N, Assendelft WJ, van der Windt DA, Hay EM, Buchbinder R, Bouter LM (2002a) Corticosteroid injections for lateral epicondylitis: a systematic review. *Pain* **96**(1–2): 23–40.

Smidt N, van der Windt DA, Assendelft WJ, Mourits AJ, Deville WL, de Winter AF et al. (2002b) Interobserver reproducibility of the assessment of severity of complaints, grip strength, and pressure pain threshold in patients with lateral epicondylitis. *Archives of Physical Medicine and Rehabilitation* **83**(8): 1145–1150.

Solway S, Beaton DE, McConnell S, Bombardier C (2002) *The Dash Outcome Measure User's Manual* (2 edn) Toronto, Canada: Institute for Work and Health.

Stahl S, Yarnitsky D, Volpin G, Fried A (1996) [Conservative therapy in carpal tunnel syndrome]. *Harefuah* **130**(4): 241–243.

Stevens JC, Sun S, Beard CM, O'Fallon WM, Kurland LT (1988) Carpal tunnel syndrome in Rochester, Minnesota, 1961 to 1980. *Neurology* **38**: 134–138.

Stratford P, Levy DR, Gauldie S, Levy K, Miseferi D (1987) Extensor carpi radialis tendonitis: a validation of selected outcome measures. *Physiotherapy Canada* **39**(4): 250–255.

Stratford PW, Levy DR, Gowland C (1993) Evaluative properties of measures used to assess patients with lateral epicondylitis at the elbow. *Physiotherapy Canada* **45**(3): 160–164.

Stratford PW, Norman GR, McIntosh JM (1989) Generalizability of grip strength measurements in patients with tennis elbow. *Physical Therapy* **69**(4): 276–281.

Struijs PA, Smidt N, Arola H, Dijk CN, Buchbinder R, Assendelft WJ (2002) Orthotic devices for the treatment of tennis elbow. *Cochrane Library* **1** http://www.thecochranelibrary.com CD001821.

Trudel D, Duley J, Zastrow I, Kerr EW, Davidson R, MacDermid JC (2004) Rehabilitation for patients with lateral epicondylitis: a systematic review. *Journal of Hand Therapy* **17**(2): 243–266.

Tuzuner S, Ozkaynak S, Acikbas C, Yildirim A (2004) Median nerve excursion during endoscopic carpal tunnel release. *Neurosurgery* **54**(5): 1155–1160.

van der Windt DA, van der Heijden GJ, van den Berg SG, ter Riet G, de Winter AF, Bouter LM (1999) Ultrasound therapy for musculoskeletal disorders: a systematic review. *Pain* **81**(3): 257–271.

Van Eerd D, Beaton D, Cole D, Lucas J, Hogg-Johnson S, Bombardier C (2003) Classification systems for upper-limb musculoskeletal disorders in workers: a review of the literature. *Journal of Clinical Epidemiology* **56**(10): 925–936.

Walters C, Rice V (2002) An evaluation of provocative testing in the diagnosis of carpal tunnel syndrome. *Military Medicine* **167**(8): 647–652.

Warren N (2001) Work stress and musculoskeletal disorder etiology: the relative roles of psychosocial and physical risk factors. *WORK: a journal of prevention, assessment & rehabilitation* **17**(3): 221–234.

Werner RA, Franzblau A, Gell N (2005) Randomized controlled trial of nocturnal splinting for active workers with symptoms of carpal tunnel syndrome. *Archives of Physical Medicine and Rehabilitation* **86**(1): 1–7.

Wright TW, Glowczewskie F, Wheeler D, Miller G, Cowin D (1996) Excursion and strain of the median nerve. *Journal of Bone and Joint Surgery, American Volume* **78A**(12): 1897–1903.

Wuori JL, Overend TJ, Kramer JF, MacDermid J (1998) Strength and pain measures associated with lateral epicondylitis bracing. *Archives of Physical Medicine and Rehabilitation* **79**(7): 832–837.

Zakaria D (2004) Rates of carpal tunnel syndrome, epicondylitis, and rotator cuff claims in Ontario workers during 1997. *Chronic Diseases in Canada* **25**(2): 32–39.

VI Orthopaedic

10 Physiotherapy Rehabilitation Following Primary Total Knee Arthroplasty

JUSTINE NAYLOR, ALISON HARMER AND RICHARD WALKER

CASE REPORT

Mrs JM, a 70 year old female, presented pre-operatively with severe tri-compartmental osteoarthritis (OA) of her right knee. On examination, she was obese (Body Mass Index (BMI) 30.8), walked with a varus thrust and a marked limp on the right, and used a walking stick. Her gait, lower limb strength, and range of motion (ROM) profiles were as follows:

- Gait speed:
  - Timed up-and-go (TUG) – 15 seconds.
  - Timed 15-m walk – 21 seconds (0.71 m/s).
  - 6-min. Walk Test (6 MWT), 322m, limited by knee pain (right > left).
- Isometric strength at $90°$:
  - Knee extensors: Right, 106 Newtons; Left, 150 Newtons.
  - Knee flexors: Right, 58 Newtons; Left, 100 Newtons.
- Knee range of motion (ROM) (passive, supine):
  - Right $= -10°$ to $100°$; Left $= -5°$ to $105°$.

Symptomatically, Mrs JM reported high pain (13/20), stiffness (5.8/5), and difficulty (45.5/68) scores on the WOMAC[1] subscales, and poor bodily pain (30/100) and physical function (26.6/100) scores on the SF-36[2] domains.

In terms of Mrs JM's medical history, she reported bilateral knee OA (right > left) of idiopathic origin of eight year's duration. She suffered from hypertension (which was controlled), ischaemic heart disease (IHD), and demonstrated poorly controlled type 2 diabetes mellitus (HbA_{1c} (glycosylated haemoglobin) 8.2 %) of seven years' duration. Consequently, her American Society of Anesthesiologists (ASA) anaesthetic risk score was estimated as II. Consequent to her multiple co-morbidity status, her

---

[1] Western Ontario & MacMaster Universities Osteoarthritis Index (low scores indicating better status).

[2] Medical Outcome Study, Short Form-36 Health related quality of life scale (high scores indicating better status).

Recent Advances in Physiotherapy. Edited by C. Partridge
© 2007 John Wiley & Sons, Ltd

medication use was extensive; for her pain management in particular, a poly-pharmacy approach was evident:

- Carvedilol, 25 mg daily.
- Glyceryl trinitrate, patch 25 mg daily.
- Metformin, 1 g bd.
- Paracetamol, prn.
- Celecoxib, 200 mg daily.
- Glucosamine and chondroitin sulphate.
- Her haemoglobin concentration (Hb) was noted to be 139 g/l.
- As part of routine anaesthetic work-up.

Socially, Mrs JM lived with her spouse in a house with 18 stairs. She had ceased recreational lawn bowls six months prior to her presentation due to pain and giving way in her right leg. She was a pensioner, reporting a low income level throughout her family life, and the highest level of education attained was primary (elementary) level.

INTRODUCTION

The benefits of total knee arthroplasty (TKA) for the individual with arthritis are perceived relatively quickly (usually within three to six months) and are generally pluralistic, including improvements in pain, ROM, knee stability, mobility, function, and health-related quality of life (HRQoL) (Aarons et al. 1996 **A**; Ethgen et al. 2004 **A**; Fortin et al. 2002 **A**; March et al. 1999 **A**; March et al. 2004 **A**; McAuley et al. 2002 **A**; Naylor et al. 2006a **A**; Pierson et al. 2003 **A**; Salmon et al. 2001 **A**; Van Essen et al. 1998 **A**). Consequently, TKA is estimated to be a highly cost-effective treatment option for severe arthritis (Segal et al. 2004 **A**). Largely ignored in cost-benefit calculations, however, are the costs associated with ongoing (post-acute care) rehabilitation. Such costs can indirectly be appreciated via the findings of March et al. (2004 **A**), who reported that the average number of out-patient physiotherapy visits by primary TKA patients was 10 in the first post-operative year, exceeding the average number of patient visits to any other health professional. This, of course, was in addition to any acute in-patient rehabilitation provided during the in-patient period (an average of 12 days) and, for many (33 %), treatment in a rehabilitation facility. We anticipate that the findings by March et al. are readily generalised as we have observed that referral to ongoing physiotherapy post-TKA is fairly routine in Australia, with out-patient based treatment predominating (Naylor et al. 2006b **A**). Our findings, obtained through a nationwide survey of TKA rehabilitation providers, echo earlier observations by Lingard et al. (2000 **A**), who reported the frequent utilisation of ongoing physiotherapy post-TKA in the UK, Australia and the US, with the latter tending to rely more on in-patient services. Given that the numbers of TKA procedures have doubled in these same countries over the last decade (Australian Orthopaedic Association National Joint Replacement Registry 2004 **A**; Dixon et al. 2004 **A**; Skinner et al. 2003 **A**), the volumes of patients potentially requiring ongoing rehabilitation to supplement surgery must also have increased. Anecdotally, in Australia at least, there is a perception that the increased surgical throughput has not been accompanied by increases or appropriate increases in the availability of downstream (ward-based and rehabilitative) resources. This must translate at some point into a time-squeeze at the therapist-patient interface and access-block for rehabilitation services. For these reasons, the need to understand the costs and benefits of rehabilitation should be an urgent priority for health systems worldwide.

Osteoarthritis (OA), the leading precipitant for TKA, is associated with significant loss of lower limb muscle strength (Fransen et al. 2003 **A**; Gur et al. 2002 **A**), walking speed (Gur et al. 2002 **A**; Lamb & Frost 2003 **A**), and function (Fransen et al. 2001 **A**). Exercise programmes involving patients with OA have repeatedly been shown to elicit significant yet small improvements in these parameters within relatively short time frames (for example, at two months) (see reviews by Bischoff & Roos 2003 **R**; Fransen et al. 2001 **R**). In contrast, TKA – a procedure typically reserved for recalcitrant arthritis – does not guarantee immediate improvements in these same parameters. Though significant improvements do occur early, several cross-sectional (Berth et al. 2002 **A**; Mizner et al. 2003 **A**; Walsh et al. 1998 **A**) and longitudinal (Benedetti et al. 2003 **A**; Lamb & Frost 2003 **A**; Lorentzen et al. 1999 **A**; Ouellet & Moffet 2002 **A**; Salmon et al. 2001 **A**) studies reveal shortfalls in gait, strength, and quality of life, compared to age-matched controls, several months to years after surgery. The argument for ongoing rehabilitation following TKA, therefore, is based on the following related contentions:

• That age-predicted norms for muscle function, gait patterns, and physical activity levels are not spontaneously or completely achieved post-surgery, and;
• That short-term exposure to prescribed interventions or physical activities will facilitate more complete recovery.

Given that the provision of acute and ongoing physiotherapeutic rehabilitation appears to be standard care across several countries, it is staggering to realise that the evidence-base which underpins rehabilitation in this area is tenuous. While there are considerable bodies of work supporting some, but not all, physiotherapeutic interventions in the acute ward phase, there is comparatively little evidence to support the various modes of ongoing rehabilitation offered either in the community or in rehabilitation wards. The trials that have been conducted (Frost et al. 2002 **A**; Kramer et al. 2003 **A**; Moffet et al. 2004; Rajan et al. 2004 **A**) all compared one mode of ongoing physiotherapy to another and did not include a true non-interventional control. Thus, the contribution of rehabilitation per se to the overall recovery process is uncertain. The lack of definitive evidence is problematic for policy makers worldwide, as health service providers are increasingly required to justify the high costs of health care, while the demand for services (in this case, rehabilitation) is increasing through sheer volume alone. Furthermore, the lack of evidence is problematic at the coalface, given that variation in practice is likely to be (Roos 2003 **C**), and has been observed to be (Naylor et al. 2006b **A**), the rule, further undermining our capacity to identify best practice.

This chapter addresses questions concerning the efficacy of various acute physiotherapeutic interventions and longer-term rehabilitative strategies Mrs JM may be exposed to through her journey of recovery. Questions concerning the impact of prosthesis type or specific surgical choices on the potential to rehabilitate or the mode of rehabilitation required are also briefly addressed. Mrs JM presents fairly typically for an elderly patient awaiting TKA for severe knee OA (Ackerman et al. 2005 A; Bozic et al. 2005 A; Heck et al. 1998 A; March et al. 2004 A; Mizner et al. 2003 A; Naylor et al. 2006a A; Ouellet & Moffet 2002 A). Notably, the measured variables are frequently utilised and recommended for the evaluation of OA and TKA (Bellamy et al. 1988 A; Ethgen et al. 2004 R; Fransen et al. 2003 A; Kennedy et al. 2005 A; March et al. 1999 A; March et al. 2004 A; Ouellet & Moffet 2002 A; Petterson et al. 2003 A). Compared to norm data or age-matched controls (see Table 10.1), the patient presents with severely compromised physical function, walking speed, range of motion, lower limb muscle strength, and HRQoL. The reported daily consumption of analgesic and anti-inflammatory medications is consistent with the high pain scores, and the use of a walking aid is somewhat typical for degenerative joint disease. It is important to note that our own experiences indicate the analgesic, anti-inflammatory, and walking aid profiles are not, in isolation, reliable

Table 10.1. Normative or age-matched physical and health-related quality of life data

	Australian Norm Data	Age-Matched Control Data
Physical Function		
SF-36 Physical Function	65.2[1]	—
WOMAC Physical Function	NA	—
Walking Speeds		
Timed up-and-go (sec)	—	8–11[2,3,4]
15-m walk (m/sec)	—	1.33–1.84[2,5]
6-minute walk (m)	—	448[2]
Isometric Muscle Strength		
Knee Extensors (N)	—	225 (sd 49)[6]
Knee Flexors (N)	—	139 (36)[6]
Health-Related Quality of life		
SF-36 General Health	64.1	—
SF-36 Vitality	60	—
SF-36 Mental Health	75.3	—
Knee Range of Motion		
Total	—	143°[4]
Pain Scores		
SF-36 Bodily Pain	69	—
WOMAC Pain	NA	—

Legend: [1]National Health Survey SF-36 Population Norms, ABS 1995 (unstandardised mean scores, female); [2]Steffen et al. 2002 A; [3]Ouellet & Moffet 2002 A; [4]Shumway-Cook et al. 2000 A; [5]Walsh et al. 1998 A; [6]Fransen et al. 2003 A; NA = not available at time of publication (Australian data). Normative data from large population sets are provided where available; otherwise, age-matched data, sourced from relevant osteoarthritis or knee replacement trials, are cited.

indicators of severity or improvement, as behavioural factors greatly influence their use.

The patient's co-morbidity profile is also typical for this patient population, with hypertension in particular being the most common co-morbidity observed in several TKA cohorts (Denis et al. 2006 **A**; Moffet et al. 2004 **A**; Naylor et al. 2005 **A**). Additionally, some physiological limitation is qualitatively suggested by the ASA score, again not atypical of TKA recipients (Bozic et al. 2005 **A**; Naylor et al. 2005a **A**; Pearson et al. 2000 **A**). Given the self-exertion nature of many rehabilitation interventions, recognising the physiological limitations imposed by concurrent illnesses is an essential consideration in any rehabilitation programme. Likewise, the socioeconomic factors, highlighted as low income and education levels, are associated with poorer pre-operative function (Ackerman et al. 2005 **A**) and some post-surgical outcomes (Fortin et al. 1999 **A**). For the therapist, these factors become relevant when setting realistic long-term patient goals and when benchmarking rehabilitation outcomes between surgical units.

REHABILITATION IN THE ACUTE PHASE

OPERATIVE HISTORY AND ACUTE POST-OPERATIVE PRESENTATION

Relevant operative details:

- General anaesthetic + femoral and sciatic nerve blocks.
- Tri-compartmental primary TKA.
- Cemented femoral, tibial, and patella components.
- Fixed-bearing, increased congruency, polyethylene bearing.
- Posterior cruciate ligament (PCL) sacrificed.
- Release of medial collateral ligament.
- Anterior cruciate ligament (ACL) removed.
- Intra-articular low suction wound drain in situ.

Presentation 18 hrs post-op (Day 1):

- Symptoms:
  - Reporting 2/10 pain on visual analogue scale, using patient-controlled analgesia c/o numbness and lack of movement in foot.
- Mobility:
  - In bed, awaiting assessment by physiotherapist.
- ROM:
  - Start flexion, $-10°$.
  - End flexion, $60°$.
  - Restricted by oedema and crepe bandaging.
  - Quadriceps lag, $15°$.

- Vital observations:
  - BP 110/70 (normally 130/80).
  - HR 95–100.
  - RR 18.
  - SpO_2 97 % (3 L/min. O_2, nasal prongs).
- Blood results:
  - Hb 105 g/l.
  - Blood glucose level (BGL) 7.7 $mmol\cdot l^{-1}$.
- Other medication:
  - Anti-hypertensives and metformin withheld.
  - Twice daily protaphane, with top up sliding scale to maintain blood glucose control.[3]

GENERAL PRINCIPLES

Rehabilitation in the acute phase is largely directed towards the minimisation of the effects of surgical trauma and rendering the patient safe for discharge. The rehabilitative strategies include the use of modalities and techniques to reduce intra- and extra-articular oedema, improve or maintain knee ROM, offset the adverse effects of bed rest, and assist independent ambulation. With respect to the determination of discharge readiness, it is recognised that some surgical units specify a minimum flexion ROM before a patient is deemed fit (Ganz & Benick 2004 **Abstract**), while others rely more on the level of function achieved (Munin et al. 1998 **A**; Naylor et al. 2006b **A**). Though speculative, the latter approach may have evolved secondary to an ever-present need to maintain patient flow in order to keep wait lists in check. In this context, the need to achieve specific physical milestones, such as a minimum flexion requirement, becomes less urgent (Benick et al. 2004 **Abstract**). It is also recognised that the threshold for discharging patients to an in-patient rehabilitation unit may differ between surgical units, with a lower threshold likely in the private market.

The nature and timing of acute care rehabilitation has also been altered over the last 10 years via the introduction of specific multi-disciplinary care pathways (protocols). Such pathways have procured impressive (up to 50 %) decreases in acute length of stay (LOS) (Brunenberg et al. 2005 **A**; Dowsey et al. 1998 **A**; Munin et al. 1998 **A**; Pearson et al. 2000 **A**; Wang et al. 1997 **A**), which must inevitably impact on the goals of rehabilitation, as the therapist-patient interface has contracted considerably at ward level. Finally, central to effective rehabilitation both now and in the longer-term, is good pain management. It is beyond the scope of this chapter to review the evolution of pain management in this context, however; suffice it to say that physiotherapists act as barometers of good pain control in their estimation of whether a patient can engage in their rehabilitation effectively.

---

[3] Additionally, referral to an endocrinologist was initiated on admission, and the recommendation was to add $1/2$ 80 mg tab of gliclazide twice daily once metformin is recommenced, with the option to increase to 80 mg twice daily if needed (i.e. if HbA_{1c} remains high).

The sources of evidence reviewed for specific rehabilitative interventions in the acute phase consisted of RCTs and systematic reviews. In order to identify the relevant literature, the following combinations of terms were used in an electronic literature search of MEDLINE, CINAHL and EMBASE:

Arthroplasty, knee, Cryotherapy.
Arthroplasty, knee, CPM.
Arthroplasty, knee, walking aid progression.
Arthroplasty, knee, exercises.

Studies were considered appropriate if the subjects had undergone primary TKA, were randomised to receive the treatment(s) under investigation, and the treatment(s) was (were) conducted in the acute in-hospital phase. In cases where a systematic review existed for a given intervention, this predominantly formed the basis for the review, to avoid duplication. Studies focusing on multi-disciplinary and multi-faceted clinical pathways were generally not included. Only studies written in English were reviewed. This review does not include the effects of pre-operative programmes on outcomes. For these, the following reviews are recommended: Ackerman et al. 2004 **R**; McDonald et al. 2004 **R**.

QUESTION 1

Does cryotherapy work?

External cooling of the knee surfaces has been shown, in the absence of haemarthrosis, to lower intra-articular temperatures in humans by 2.7–5 °C (Martin et al. 2002 **A**). This, together with the local effects of cold therapy on neural and vascular function, presumably motivates the use of cryotherapy post-TKA for the purposes of reducing pain and swelling. The use of cryotherapy has been observed to be inconsistent in the acute phase following TKA, in terms of both the factors governing its application (Barry et al. 2003) and whether it is utilised at all (Naylor et al. 2005 **A**, 2006b **A**). To date, cryotherapy post-TKA has not been systematically reviewed, but several RCTs have been conducted (Gibbons et al. 2001 **A**; Healy et al. 1994 **A**; Ivey et al. 1994 **A**; Scarcella & Cohn 1995 **A**; Smith et al. 2002 **A**; Webb et al. 1998 **A**). Only one study (Webb et al. 1998 **A**), comparing cold compression to a non-interventional control, observed significantly less blood transfusions, analgesic consumption, and pain with cold therapy. Of course, the contribution made by the compression component could not be differentiated in this study. Of note, despite the pain relief and blood loss benefits, no differences in ROM acutely or at 12 weeks were observed. For the majority of the remaining studies in this area, no or minor differences have been observed between those receiving and not receiving early cryotherapy on several outcomes, including LOS, transfusion needs, swelling, ROM, pain, and analgesic use. Having said this, the interpretation of the impact of cryotherapy in these studies is clouded by comparisons with alternative treatments (such as compression bandaging or alternative cold therapy) (Gibbons et al. 2001 **A**; Healy et al. 1994 **A**; Smith et al.

2002 **A**) rather than comparisons with true non-interventional controls. Healy and colleagues (1994 **A**) compared cryotherapy to ice packs. Smith et al. (2002 **A**) used cold therapy in both groups after 24 hours. Scarcella and Cohn (1995 **A**), with their sample of 24 TKA patients, were not likely to have had sufficient power to detect differences between their groups when others (Smith et al. 2002 **A**) have required a sample of 80 for the same outcome variables. Finally, Gibbons et al. (2001 **A**) did not account for possible gender differences in Hb levels between the treatment and control groups, which themselves differed in their gender profile. This may have explained why cold compression was not associated with a lower transfusion requirement in this study despite being associated with smaller post-operative blood losses. Even with the lack of irrefutable evidence demonstrating that there is no additional benefit from cryotherapy, various authors (Healy et al. 1994 **A**; Smith et al. 2002 **A**) have concluded that its costs outweigh its benefits and that compression is preferred in light of this. We conclude that although at this stage it would appear that cryotherapy offers no additional benefits beyond those which could be achieved with compression alone, the methodological limitations of the majority of studies conducted render this issue unresolved.

Regarding Mrs JM, the available evidence does not strongly support or refute the use of cryotherapy, nor is it clear whether compression bandaging alone is superior to it. Thus, the therapist would be justified in trying either. Ideally these modalities would be applied both before and after physiotherapy; at the very least, pain, oedema and ROM should be monitored pre- and post-application. However, Mrs JM's initial numbness – presumed at this stage to be a hangover from her intra-operative regional anaesthetic – may delay the commencement of ice therapy. Of course, neural deficits beyond 24 hours will need to be differentiated from possible chronic loss due to diabetic neuropathy. Though speculative at this point, the presence of the haemarthrosis following TKA may undermine the impact of external ice applications, rendering the effects of compression bandaging more important.

QUESTION 2

Does continuous passive motion work?

Continuous passive motion (CPM), like cryotherapy, is an adjunctive rehabilitation tool intended to decrease swelling and haemarthrosis, and enhance soft tissue healing and joint ROM (Milne et al. 2003 **A**). In contrast to cryotherapy, however, CPM has been subject to many RCTS involving TKA recipients (n = 59), one Cochrane review (Milne et al. 2003 **A**), and one qualitative review (Lachiewicz 2000 **R**). Thus, more definitive conclusions can be drawn regarding its effectiveness.

Milne et al. (2003 **A**), based on a meta-analysis, concluded that CPM combined with standard physiotherapy was associated with a small increase in flexion ROM at two weeks (4.3° weighted mean difference (WMD[4])), decreased LOS (0.69 days

---

[4] WMD: difference between control and treatment group is weighted by the inverse of the variance.

WMD), and a decreased risk of manipulation within the first month (relative risk 0.12). CPM was not found to improve passive ROM. The authors did conclude, however, that information and protocol biases were present in the review due to inadequate reporting of some variables (for example, whether ROM was passive or active) and inconsistent protocols (for example, pain relief and pre-operative education) across studies. Information on ideal dose and application could not be derived. In light of these facts, the authors recommended that the potential benefits of CPM be weighed against the possible increased costs and inconvenience, and that more research be conducted to determine the optimum treatment parameters. Not included in the analyses were the effects of CPM on midline wound healing, bleeding overall, and hospital costs. These have been shown to be a concern in some trials (Lachiewicz 2000 **R**).

Since the publication of the meta-analysis by Milne and colleagues, only one other RCT has been conducted in TKA patients. Denis et al. (2006 **A**) did not observe any differences in discharge ($\sim$ eight days post) ROM, LOS, WOMAC function, and TUG times between those treated with conventional physiotherapy plus 35 or 120 minutes of CPM daily, and those receiving conventional physiotherapy only. With the exception of LOS, these results confirm the conclusions of the aforementioned meta-analysis. It is unfortunate, however, that the number of manipulations post-discharge was not monitored given that this is perhaps the most clinically relevant outcome concerning CPM.

In terms of current clinical practice, we observed that CPM does not appear to be in routine use in Australia (Naylor et al. 2006b **A**). Whether this is the case elsewhere is unknown as there are no other survey data concerning this. We also observed in our unit, where CPM was routinely prescribed, that only 40 % of patients received it (Naylor et al. 2005 **A**). Protocol deviance was explained by a combination of lack of awareness of the protocol by rotating physiotherapists, and their perceived lack of need – the latter possibly explained by the fact that functionality and not ROM primarily determines discharge at our unit. At this point in time, our CPM practices, together with our pain relief and pre-operative education policies, are under review, as the number of manipulations performed within six months of surgery has increased in recent times.

Regarding Mrs JM, in view of the risk of manipulation alone, CPM should be initiated at least once per day for several hours during bed rest periods. This recommendation ideally applies to units where CPM is readily available and where medical and nursing staff can apply it. Though speculative, CPM may be of particular benefit to Mrs JM given her poorly controlled diabetes (evidenced by the elevated HbA_{1c} of 8.2 %; non-diabetic range 3–6 %). Glycosylation (permanent protein modification by glucose) of collagen or elastin as a result of persistently high BGL may result in tissue stiffness (Paul & Bailey 1996 **B**), hence Mrs JM may be at a greater risk of manipulation.[5]

---

[5] 22 % of patients presenting for manipulation under anaesthesia for frozen shoulders had diabetes (Hamdan & Al-Essa 2003).

QUESTION 3

What is the evidence for exercise and early ambulation to improve ROM, decrease length of stay, and prevent deep venous thrombosis?

Only one study has compared the outcomes of patients who received formal knee flexion exercises in addition to standardised physiotherapy with those who received standardised physiotherapy only (Ganz & Ranawat 2004 **Abstract**). Though formal knee flexion exercises were associated with greater active knee flexion at one week, this did not translate into any functional differences (such as stair ambulation or use of aids) or shorter LOS. At three and 12 months, there were no differences in active knee flexion. No studies were found focusing on active knee extension. Despite the lack of evidence in support of specific active exercises, we have observed the prescription of lower limb exercises in the acute stage to be routine in Australia (Naylor et al. 2006b **A**). This notwithstanding, as there does not appear to be any routine case to suggest active exercises are detrimental in this patient group, we find no reason for not including them in the therapy repertoire.

Similarly to active exercises, the arguments for early ambulation post-TKA rest largely on the desire to minimise the well-known adverse effects of bed rest and to accelerate discharge from hospital. To our knowledge, only one RCT has been conducted (Munin et al. 1998 **A**) which highlights the specific benefits of early rehabilitation, including early ambulation (commencing Day Three versus Day Seven), on LOS, functional performance, and Deep Vein Thrombosis rate. Though the specific contribution attributable to early ambulation alone cannot be reliably estimated, the absence of evidence to the contrary suggests protocols aimed at early ambulation are desirable. We do qualify this statement, however, in that we recommend an assessment of the patient's medical stability (including blood pressure, heart rate and rhythm, BGL, oxygen saturation levels, and Hb) precedes any physiotherapy intervention.

Regarding Mrs JM, her lower limb neural deficit will preclude ambulation and some bed exercises until it resolves. A combination of closed- and open-chain isometric, concentric, and eccentric exercises will be prescribed for the flexor and extensor muscle groups in her lower limbs. Ambulation will commence after removal of the wound drains. Her cardiovascular history necessitates close monitoring of her vital signs prior to her participating in any exercise, however. Her low Hb is typical at this stage, given the acute blood losses (mean 608 mls) associated with the surgery (Naylor et al. 2005 **A**), and, at her current level, does not warrant a transfusion (NH&MRC & ASBT 2001 **A**).

QUESTION 4

What evidence guides walking aid progression?

The literature search yielded no RCTs investigating the optimal rate of walking aid progression. We are aware of surgical units that dictate the rate of progression according to the presence or absence of cement. In our unit, all patients are progressed and discharged on crutches, with instructions to weight-bear as tolerated unless otherwise indicated. It is not clear at this stage whether the rate of progression onto a walking stick or to complete independence from walking aids is a concern for long-term prosthesis stability, the restoration of normal gait patterns, or the evolution of back pain.

QUESTION 5

Does electrical stimulation work?

The electrical stimulation of the knee extensor muscles post-TKA is based on the premise that voluntary activation is not sufficient to restore strength (Avramidis et al. 2003 **A**). Three studies were identified that randomised the use of electrical stimulation to the vastus medialis or quadriceps femoris during CPM, commencing in the acute period and given alongside a standardised physiotherapy programme. Gotlin et al. (1994 **A**) and Haug and Wood (1988 **A**) observed that patients receiving two to three hours of muscle stimulation daily until discharge experienced less extensor lag and shorter LOS. In a longer-term study, Avramidis et al. (2003 **A**) observed that patients receiving electrical stimulation for two hours twice daily from the second post-operative day for six weeks, attained a faster walking speed at six weeks, and this effect carried over until the 12th week. The authors concluded that the greater walk speed was a consequence of more rapid quadriceps recovery and, as such, a greater ability to participate in exercise. It should be noted that the control group did not receive any standardised physiotherapy post-discharge. The addition of a third group that received standardised physiotherapy for six weeks would have helped to clarify whether electrical stimulation was superior to or simply a replacement for voluntary muscle activation. While the use of electrical stimulation looks promising, the technical and potentially cumbersome nature of the procedure, and the prerequisite for effective communication between patient and therapist for safety reasons, may have deterred widespread adoption of this treatment option.

Regarding Mrs JM, assuming availability of the device and competency of both the staff and patient in its use, intermittent neuromuscular stimulation is an appropriate rehabilitation intervention, given her quadriceps lag.

QUESTION 6

What is the evidence for hydrotherapy?

No RCTs were identified concerning the efficacy of hydrotherapy post-TKA. We recognise that it is a treatment option where facilities exist (Naylor et al. 2006b **A**) and that a non-randomised trial has been conducted in Germany (Erler et al. 2001 **A**). No recommendations can be made at this stage, but note that, at the very least, the integrity of the wound is paramount for hydrotherapy to be considered a viable option.

REHABILITATION IN THE POST-DISCHARGE PHASE

GENERAL PRINCIPLES

The post-discharge phase of rehabilitation commences after discharge from an acute care facility. Goals of rehabilitation in the earlier post-discharge phase focus upon increasing the level of independence of the patient, which may include weaning off a walking aid, maintaining and improving knee joint ROM, controlling or reducing residual oedema, increasing muscle strength and endurance, and gradual return to work and leisure activities. In the later phase of post-discharge rehabilitation, goals include further improvement of muscle strength and endurance, improvement of cardiovascular fitness, and full return to work and leisure activities.

The sources of evidence reviewed for specific rehabilitative interventions in the post-acute phase consisted of RCTs and systematic reviews. In order to identify the relevant literature, the following combinations of terms were used in an electronic literature search of MEDLINE, CINAHL and EMBASE:

Total knee replacement, with subject headings: arthroplasty, replacement, knee; knee prosthesis; TKR.
Rehabilitation, with all subject headings.
Physiotherapy, with subject headings: exercise therapy; orthopedics; physical therapy (specialty); physiotherapy.

The initial literature search yielded 230 studies. For the present review, studies were only included if the subjects had undergone primary TKA, were randomised to receive the treatment(s) under investigation, and the treatment(s) was (were) conducted in the post-acute phase. Only studies written in English were reviewed. Only five trials satisfied these criteria, thus revealing the paucity of evidence for effects of rehabilitation in the post-acute phase. One study (Mitchell et al. 2005 **A**) included pre-operative physiotherapy in one group and was thus excluded. The remaining four trials differed markedly in their methodology and investigated the effects of out-patient physiotherapy versus home-based rehabilitation (Kramer et al. 2003 **A**; Rajan et al. 2004 **A**); traditional versus functional home-based exercise (Frost et al. 2002 **A**); and intensive versus usual care treatment (Moffet et al. 2004 **A**). Due to the limited number of studies identified and the holistic nature of the physiotherapy programmes described, it was not possible to examine the effect of a single treatment component in the post-acute phase. In addition to the five reports of randomised trials, one recent review that presented current evidence from experts on knee and hip arthroplasty was identified (Jones et al. 2005 **R**).

QUESTION 7

What is the evidence supporting early post-discharge rehabilitation?

Three RCTs have examined the effects of physiotherapy provided in the early post-discharge phase of rehabilitation; that is, commencing immediately after discharge from acute care. Kramer et al. (2003 **A**) investigated effects of clinic- versus home-based rehabilitation. All patients were provided with advice on knee management and were prescribed home strengthening and ROM exercises, the basic form of which they were taught during the acute in-patient period. The home-based group received weekly phone calls from a physiotherapist, whereas patients in the clinic-based group attended the clinic once or twice weekly until three months post-operation. At three and 12 months post-operation there was no difference between groups on any outcome measures, which included WOMAC total and pain and function subscales, SF-36 total, knee flexion range, 30-second stair test, and 6MWT. Similarly, another study (Rajan et al. 2004 **A**) found no additional benefit of out-patient physiotherapy compared with a home exercise programme at three, six, or 12 months; however, there was no description of the physiotherapy interventions, and the only outcome measure reported was knee flexion range. Provided that sufficient knee range is available for performance of ADL, this outcome measure is a poor sole criterion upon which to judge treatment efficacy in the post-acute phase. Frost et al. (2002 **A**) compared two home-based programmes – usual care (for example, ROM exercises, quadriceps, and hamstrings strengthening) versus functional exercises (rising from a chair, lifting the leg onto a step, and walking) – that commenced immediately after hospital discharge. At the one-year follow-up assessment there was no difference between groups in 10 m walking speed, pain, knee flexion range, or leg extensor power.

All three of the above studies used intention-to-treat analysis; one study employed therapist blinding (Frost et al. 2002 **A**) and another, partial blinding (Kramer et al. 2003 **A**), and subjects were randomly allocated to groups. Losses to follow-up were 3 % (Rajan et al. 2004 **A**), 23 % (Kramer et al. 2003 **A**), and 43 % (Frost et al. 2002 **A**), and all studies described reasons for drop-out. Very few adverse events occurred using the exercises prescribed in these studies. According to the principles of evidence-based practice (Herbert et al. 2005 **A/R**), the Physiotherapy Evidence Database (PEDro) assigned the following scores to each of the studies: Frost et al. 6/10; Kramer et al. 6/10; and Rajan et al. 7/10; indicating that these studies all provide a moderate level of evidence. It can be concluded that patient outcomes one year post-TKA are not affected by location of rehabilitation delivery (out-patient physiotherapy clinic versus home) or type of exercise (usual versus functional). However, loss to follow-up may be affected by the level of supervision provided by the physiotherapist (out-patient attendance or phone call monitoring versus no monitoring). Larger trials, which provide a greater power to detect small differences in outcome measures, may necessitate revision of these conclusions. Patient outcomes at one year post-TKA indicate that although significant improvements were evident compared to before surgery, there is still a residual level of pain, disability, and loss of knee flexion range; and that patients only just attain the lower limits of age-matched normal function, for example walking speed. A lack of sufficient exercise intensity during rehabilitation may partly contribute to these shortfalls in recovery, but it was not possible to calculate exercise dosage from these trials since exercise intensity was largely patient determined or else it was not described.

QUESTION 8

What is the evidence supporting later post-discharge rehabilitation?

One RCT only (Moffet et al. 2004 **A**) has examined the effect of commencing rehabilitation later in the post-discharge phase. Ability to exercise in this stage would be anticipated to be greater than in the early post-acute phase, when anaemia, pain, oedema, and residual effects of anaesthesia can cause significant limitation. Moffet et al. (2004 **A**) employed intention-to-treat analysis: blinding of evaluators; random allocation of subjects; and had only ∼10 % loss to follow-up, with all drop-outs being described, thus providing a moderate to strong level of evidence (PEDro score 7/10). Two months after TKA, patients were randomised to either usual care (strength training, ROM exercise, ice, gait retraining; 26 % also received home visits) or to an intensive 12-week supervised physiotherapy programme, which also included the usual care components. Intensive sessions included strength (for example, maximal isometric contractions of quadriceps and hamstrings; functional exercises such as sit-to-stand and stairs) and endurance exercise training (walking or cycling at 60–80 % of maximum predicted heart rate for up to 20 min.). Exercise intensity was progressed as required, however, while number of repetitions was reported, intensity of strength training was difficult to assess from the data provided. No adverse events from treatment occurred. At six months post-TKA, patients in the intensive exercise group had increased their 6MWT by 31 % (93 m), compared to 25 % (72 m) increase in the usual care group; a significant effect size between interventions of ∼9 %. Significant treatment effect differences of a similar magnitude were evident in the WOMAC subscales of pain, stiffness, and difficulty in performing ADL. One year after TKA, patients in the intensive group tended to have a higher 6 min. Walk Test distance ($P = 0.06$; 400 m or ∼1.1 m·s^{-1}, which placed them at the lower limit of normal for their age) than the control group (370 m; 1.03 m·s^{-1}), and both groups had similar levels of pain, stiffness, and difficulty performing tasks. This study demonstrates that more intensive rehabilitation, commenced in the later post-acute phase, results in greater improvements in walking speed at six months post-TKA (and probably also at 12 months, given the near statistical significance and relatively low subject number). Therefore, usual care physiotherapy after TKA probably provides less than optimal stimuli, and patients could likely make further significant gains if sufficiently challenged in the post-discharge rehabilitation period. Further, the authors suggest that increasing the exercise intensity and prolonging the programme may yield greater treatment effects. If so, this not only has important functional relevance for the patient, but also has implications for the progression or retardation of common co-morbidities such as hypertension and type 2 diabetes.

POST-DISCHARGE REHABILITATION FOR MRS JM

Mrs JM has similar co-morbidities (HT, diabetes, cardiac disease) and is of a similar age to the patients in the Moffet et al. (2004 **A**) study. Her scores for each of the WOMAC subscales are two- to three-fold higher than those reported at two months post-TKA, and are anticipated to improve considerably after surgery. Ideally, Mrs JM's early post-acute rehabilitation will be conducted from home; however, a retrospective review of effects of the co-morbidities of HT, diabetes, and obesity (all of which Mrs JM suffers from) in 959,839 patients after arthroplasty found that each of the co-morbidities was an independent predictor of increased post-operative complications and non-homebound discharge (Jain et al. 2005 **A**). Additionally, achievement of rehabilitation goals by Mrs JM may be slowed by the presence of OA in her left knee (unoperated). For example, progression from a walking aid to independent ambulation, or the recovery and improvement of walk speed, may be delayed by poor ipsilateral or contralateral joint dysfunction. Certainly, with respect to the latter, our data demonstrate that 15-m walk and TUG times are slower in patients with a knee or hip replacement awaiting further surgery for other joints than in patients with knee or hip replacement who are not (Naylor et al. 2006a **A**).

Based on the evidence from the RCTs discussed above, in the early post-acute phase, Mrs JM will be prescribed an exercise programme that includes ROM and strength exercises (including functional exercise), and gait retraining; and she will receive advice regarding management of oedema and pain. Mrs JM will remain relatively anaemic (Hb 105 $g \cdot l^{-1}$) at discharge, which may result in mild fatigue, dizziness, and dyspnoea during more demanding submaximal exercise, as a consequence of lower arterial oxygen content. This, coupled with pain, oedema, and the associated muscle inhibition, will reduce the exercise intensity that Mrs JM can undertake in this early period. In addition, given Mrs JM had poor pre-operative control of her diabetes (indicated by the HbA_{1c}), she may experience more difficulty controlling her BGL in the post-acute phase consequent to reduced activity, stress, and hospitalisation. Even so, current opinion (Sigal et al. 2004 **A**) is that light- or moderate-intensity exercise should not be postponed in those with type 2 diabetes, even if BGL exceeds ~ 17 $mmol \cdot l^{-1}$ (300 $mg \cdot dl^{-1}$), unless the patient feels unwell and has urinary or blood ketones. We anticipate improved blood glucose control in this case, following the review by the endocrinologist in hospital and the consequent addition of gliclazide to Mrs JM's usual metformin. Advice from a diabetes educator and a dietician will also enhance her management. Her programme can be conducted at home, with a weekly phone call from her physiotherapist to assess her ability to complete the exercises, to advise on exercise progression, and to monitor potential complications.

In the later post-discharge phase, Mrs JM will attend out-patient physiotherapy for a more intensive programme. The programme will commence once oedema and pain have subsided; probably about six to eight weeks post-surgery, and will build upon the gains made with therapy in the acute period. Additionally, based on our recent audit of acute and short-term outcomes following TKA (Naylor et al. 2005 **A**), we anticipate that Mrs JM will have recovered to ~ 90 % of her pre-operative Hb (~ 125 $g \cdot l^{-1}$) by about the sixth week post-TKA. Given the presence of type 2 diabetes mellitus, hypertension, and IHD, current recommendation (Sigal et al. 2004 **A**) is that it would be prudent to have Mrs JM formally assessed for cardiovascular risk prior to commencing more intense exercise (if not done comprehensively pre-operatively). Following individual evaluation and exercise prescription, her programme can be undertaken in a group setting, which may be a more cost-effective way to deliver more intense, supervised rehabilitation, and may enhance motivation. The programme will include lower limb strength training, functional exercises to promote strength and balance, stretches, and either cycling or walking for local muscular and whole-body endurance. Intensity will be monitored by heart rate and rating of perceived exertion (the latter particularly, if any autonomic neuropathy is suspected or demonstrated), and the number of repetitions and sets of each strength exercise, and the load and duration of endurance exercises will be recorded. As previously stated, the current evidence does not provide sufficient detail to determine exercise dosage for resistance training; hence the following suggestions are based upon research drawn from other sources, and are subject to change when further specific evidence regarding resistance training after TKA is published. The intensity of resistance exercise will be gradually increased as tolerated, beginning with one set of 10–15 repetitions (not to fatigue) twice per week, and over a number of weeks progressing to three sets of eight repetitions at a 10 RM (repetition maximum) load up to three times per week. The latter is recommended for individuals with type 2 diabetes, to assist with improving metabolic control, for example lowering HbA_{1c} (Sigal et al. 2004 **A**) – a very desirable outcome in Mrs JM. Resistance exercise is also recommended for patients with OA; however, it is suggested that muscles should not be exercised to fatigue (American Geriatrics Society Panel on Exercise and Osteoarthritis 2001 **A**). Hence, resistance training for the left leg (knee OA) will be conducted at a lower load and not to fatigue (for example, eight to 10 repetitions at 15 RM load) and will be changed to isometric exercise if the left knee becomes unstable or acutely inflamed. Endurance exercise (walking or cycling) will be commenced at 50 % of maximum heart rate for five to 10 minutes at least every second day, and progressed as tolerated to a weekly dose of 150 minutes at 50–70 % of maximum heart rate (Sigal et al. 2004 **A**). Based on the results of Moffet et al. (2004 **A**), Mrs JM can expect to be walking ~30 % further in a 6MWT after six months; perhaps even more quickly if she has a more intense exercise programme that is continued for a longer period (depending on the degree of limitation from her left knee OA). The ability to undertake both sustained aerobic and resistance exercise is important in addressing Mrs JM's co-morbidities of obesity, type 2 diabetes, HT, and IHD, and in accomplishing a full return to her ADL (including negotiation of 18 stairs at home) and leisure activities (lawn bowls). In addition, consultation with a diabetes educator and a dietician are recommended for Mrs JM.

Thus it appears that most rehabilitation programmes finish just when the patient is becoming more capable of exercising with greater intensity. The incorporation of more challenging (more intense) exercise may address the deficits in gait speed, muscle strength, and quality of life evident several months to years after TKA (see Introduction).

Given the common occurrence of co-morbidities in patients who undergo TKA, a more protracted exercise programme, which included both strength and endurance components, would be anticipated to have important health and financial benefits. However, given the paucity of RCTs and the holistic nature of the existing post-acute physiotherapy RCTs, there is little evidence to suggest what the optimal exercise programme after TKA might comprise.

IMPACT OF SURGICAL FACTORS ON LONGER TERM RECOVERY

QUESTION 9

Do prosthesis design and surgical choice impact on rehabilitation or functional recovery?

Despite myriad investigations concerning efficacy of TKA, there is significant variation in the prostheses used and surgical decisions made. In other words, despite substantial evidence supporting the intervention, best surgical practice in this field is yet to be recognised. The need to re-align the knee to a neutral mechanical axis and balance the soft tissue is generally agreed upon; some of the issues that remain debated in the literature include cemented versus uncemented implants, the role of the posterior cruciate ligament, mobile versus fixed bearing, and whether or not to resurface the patella.

Cemented versus uncemented fixation

Mrs JM underwent a cemented TKA. Cemented TKA remains the standard to which alternative forms of fixation need to be compared (Insall et al. 1976 **A**; Jones et al. 2005 **A**; Rodriguez et al. 2001 **A**). In Australia, cemented TKA make up almost 50 % of procedures, while uncemented and hybrid implants comprise 25 % each (Australian Orthopaedic Association National Joint Replacement Registry 2004). Uncemented fixation has the theoretical advantage of osseointegration, which may have implications for longevity, infection, and future bone loss (Diduch et al. 1997 **A**), while cemented implants have a significant cost benefit. In general terms, failure of uncemented implants has been mainly on the tibial and patella surfaces. Many early designs showed pain scores that were slower to improve than in their cemented counterparts, and had higher revision rates (Duffy et al. 1998 **A**; Ritter 2001 **A**). Newer implant designs may have overcome these problems; however, long-term results are yet to be realised. To date there is no literature examining the impact of weight bearing on early and late fixation in cemented or uncemented prostheses.

Cruciate versus no cruciate

While most current prostheses sacrifice the ACL, controversy remains regarding the PCL. Some argue that preservation of the PCL aids in improving the stability, kinematics, and mechanics of the knee replacement and avoids extra bone resection (Rand 1996 **A**). Those in favour of excision argue that the PCL is not normal in arthritic knees and that its excision allows improved balancing and correction of deformity (Hirsch et al. 1994 **A**), as well as more consistent and predictable kinematics (Dennis et al. 1996 **A**; Dennis et al. 1998 **A**). Excellent clinical results have been shown with both PCL-retaining and -sacrificing TKAs. Nevertheless, there remain significant differences in the kinematics between normal and replaced knees, and much of this gait abnormality is thought to be related to cruciate deficiency. This, combined with senile muscle weakness and prolonged disability, may further reduce the ability of patients to perform activities, including rehabilitative activities, following TKA.

Two options are available to improve stability and kinematics following resection of the ACL, PCL, or both. One option is to increase the congruity of the polyethylene with anterior and posterior lips, to prevent translation of the tibia relative to the femur. The other option is for the surgeon to introduce a cam-and-post mechanism, which prevents posterior translation of the tibia relative to the femur. Mrs JM had a posterior stabilised cam-and-post type implant. It is important to recognise that neither PCL-retaining nor -substituting implants provide varus or valgus stability, and they both require intact collaterals for stability.

A recent RCT (Straw et al. 2003 **A**) examined the effect of the PCL in total knee arthroplasty. Patients were randomised to retention or excision of the PCL. There were four groups: (a) PCL retaining and standard implants; (b) PCL released and standard implants; (c) PCL excised and standard implants; (d) PCL excised and posterior substituting implants. There was no difference in groups (a), (c) and (d) with regards to pain scores, range of motion, knee scores, or functional scores. Patients in group (b), with retaining implants and a released PCL, did significantly worse than the other three groups in terms of knee scores and function. The posterior stabilised group (d) had the highest functional scores, walking distance, and stair climbing. The poorest range of motion was in group (a), suggesting tightness in flexion with PCL retention. In terms of clinical stability, posterior stabilised (d) were the most stable, while the excised group (c) were the most lax in the anteroposterior plane; this was not statistically significant, however. There was no difference between groups in terms of mediolateral stability. Follow-up averaged 3.5 years and as such the issue of long-term wear could not be examined.

Integrity of the collateral ligaments

Release of the collateral structures is required during TKA when the gaps created for the implants in flexion and extension are not rectangular. If left asymmetrical, this can lead to asymmetric forces on the medial or lateral sides of the knee and potentially cause pain, instability, poor function and early wear. Creating equal gaps requires correct bony alignment as well as appropriate release of the soft tissues. In the varus knee, medial structures tend to be tight, while in the valgus knee, it is the lateral structures that become tight. The contributing structures will depend on whether the knee is tight in an extended or flexed position. On the medial aspect, the medial collateral ligament and postero-medial capsule may require releasing to balance the knee (Whiteside 1995 **A**). On the lateral aspect, the lateral collateral, popliteal tendon, iliotibial band and capsule may need releasing for balance (Whiteside 1999 **A**). Mrs JM required release of the medial collateral ligament to balance the knee. Occasionally in severe deformity, the opposite side attenuates (for example, medial structures in a valgus knee), thus, requires attention. If so, surgical reconstruction of the ligament is performed, a more constrained form of implant is used, or both. Instability following collateral release does not occur provided that the mechanical axis of the leg has been corrected with the surgery. Ligament releases still leave peripheral attachments and other soft tissue connections, such as periosteum or capsular tissue, which allow the released ligaments to function (Whiteside 2005 **A**). Ongoing clinical instability, perhaps detected by the therapist if not reported by the patient, usually occurs in the presence of overall limb malalignment, inadequate soft tissue release, or inadvertent transection of ligamentous structures. By and large, collateral release should not impede functional recovery or rehabilitation.

Fixed versus mobile bearing implants

While fixed bearing implants yield excellent results, mobile bearing prostheses were introduced to try and improve wear characteristics, range of motion, and longevity. These implants have dual articulation, with a highly conforming articular surface between the femur and the polyethylene insert. Many designs exist and they vary in the degree of movement allowed between the polyethylene and the base plate. One long-term non-randomised study reported similar clinical and prosthesis survivorship results to fixed bearing implants (Buechel 2002 **A**), but the impact of activity per se, either early or late, was not addressed. It is tempting to speculate that, given the equivalent prosthesis survivorship across the two designs, neither activity level nor type of activity impacts on long-term functional recovery. However, non-randomised allocation of patients to the varying prosthetic designs may contribute to this; thus, RCTs are ideally needed to confirm this notion. It is also worth noting that trials subjecting the same prostheses to differing long-term in vivo mechanical loading (such as functional and exercise loads) have not been conducted.

Patella resurfacing versus non-resurfacing

Controversy remains over whether or not to resurface the patella. Mrs JM had a cemented patella resurfacing. Many of the early problems with the patellofemoral joint have been addressed by improving the characteristics of the femoral component (Andriacchi et al. 1997 **A**) and, as such, much of the older literature may not be relevant today. Ongoing anterior knee pain is the reason for considering resurfacing, while complications including patella fracture, extensor mechanism disruption, and loosening are reasons to avoid this option routinely. Resurfacing of the patella is generally agreed upon in inflammatory arthritis, patella maltracking, eburnated bone on the patella, preoperative anterior knee pain, and crystalline deposition disease (Kajino et al. 1997 **A**; Kim et al. 1999 **A**). Studies in patients with bilateral arthroplasty with only one side resurfaced have not shown significant differences (Keblish et al. 1994 **A**). While there is equivocal evidence from RCTs (Barrack et al. 2001 **A**; Wood et al. 2002 **A**), a recent review (Holt & Dennis 2003 **R**) concluded that, although patient selection is critical to the decision to resurface the patella, unresurfaced patellae deteriorate over time and secondary resurfacing is associated with greater residual patellofemoral pain. This was reiterated by Jones et al. (2005 **R**), who also concluded that patella resurfacing is likely to improve outcomes, including long-term pain-free patella function. From the therapist's perspective, knowledge of whether or not the patella was resurfaced may help explain ongoing or residual anterior knee pain, or even pain emerging within a few months to years of the TKA procedure. To our knowledge, there are no context-specific data available to guide the therapist in terms of what, if any, lower limb exercises are preferred in the presence or absence of patella resurfacing.

SUMMARY

The choices that surgeons face when undertaking TKA are manifold. Unfortunately, well constructed RCTs are not available to answer many of the debates that remain, particularly in relation to TKAs' relevance to rehabilitation. However, from a surgeon's perspective, there is little doubt that good alignment and good balance are the most important features in providing patients with a well-performing, long-lasting joint replacement. Provided these principles are adhered to, and once best practice rehabilitation is identified, we assume at this stage that post-operative physiotherapy and rehabilitation should not be substantially affected by variations in surgical hardware and technique. Having said that, note that the more cognisant the physiotherapist is of each patient's surgical particulars, the less risk there is of their doing harm, and the better placed they are to set pragmatic rehabilitation goals.

REFERENCES

Aarons H, Hall G, Hughes S, Salmon P (1996) Short-term recovery from hip and knee arthroplasty. *Journal of Bone and Joint Surgery* **78B**: 555–558.

Ackerman IN, Bennell KL (2004) Does pre-operative physiotherapy improve outcomes from lower limb joint replacement surgery? A systematic review. *Australian Journal of Physiotherapy* **50**: 25–30.

Ackerman IN, Graves SE, Wicks IP, Bennell KL, Osborne RH (2005) Severely compromised quality of life in women and those of lower socioeconomic status waiting for joint replacement surgery. *Arthritis & Rheumatism* **53**: 653–658.

American Geriatrics Society Panel on Exercise and Osteoarthritis (2001) Exercise prescription for older adults with osteoarthritis pain: consensus practice recommendations. *Journal of the American Geriatrics Society* **49**: 808–823.

Andriacchi TP, Yoder D, Conley A et al. (1997) Patellofemoral design influences function following total knee arthroplasty. *Journal of Arthoplasty* **12**: 243.

Australian Orthopaedic Association National Joint Replacement Registry (2004) *Annual Report* Adelaide: Australian Orthopaedic Association.

Avramidis K, Strike PW, Taylor PN, Swain ID (2003) Effectiveness of electric stimulation of the vastus medialis muscle in the rehabilitation of patients after total knee arthroplasty. *Archives of Physical Medicine and Rehabilitation* **84**: 1850–1853.

Bachmeier CJM, March L, Cross M, Lapsely H, Tribe K, Courtenay B et al. (2001) A comparison of outcomes in osteoarthritis patients undergoing total hip and knee replacement surgery. *Osteoarthritis and Cartilage* **9**: 137–146.

Barrack RL, Bertot AJ, Wolfe MW, Waldman DA, Milicic M, Myers L (2001) Patellar resurfacing in total knee arthroplasty: a prospective, randomized, double-blinded study with five to seven years of follow-up. *Journal of Bone and Joint Surgery* **83A**: 1376–1381.

Barry S, Wallace L, Lamb S (2003) Cryotherapy after total knee replacement: a survey of current practice. *Physiotherapy Research International* **8**: 111–120.

Bellamy N, Buchanan W, Goldsmith C, Campbell J, Stitt L (1988) Validation study of WOMAC: a health status instrument for measuring clinically-important patient-relevant outcomes following total hip or knee arthroplasty in osteoarthritis. *Journal of Orthopaedic Rheumatology* **1**: 95–108.

Benedetti MG, Catani F, Bilotta TW, Marcacci M, Mariani E, Giannini S (2003) Muscle activation pattern and gait biomechanics after total knee replacement. *Clinical Biomechanics* **18**: 871–876.

Benick RA, Backus SI, Kroll MA, Ganz SB, MacKenzie CR (2004) Knee flexion and functional ambulatory status following unilateral total knee arthroplasty. *Topics in Geriatric Rehabilitation* **20**: 308.

Berth A, Urbach D, Awiszus F (2002) Improvement of voluntary quadriceps muscle activation after total knee arthroplasty. *Archives of Physical Medicine and Rehabilitation* **83**: 1432–1436.

Bischoff HA, Roos EM (2003) Effectiveness and safety of strengthening, aerobic, and coordination exercises for patients with osteoarthritis. *Current Opinion in Rheumatology* **15**: 141–144.

Bozic KJ, Durbhakula S, Berry DJ, Naessens JM, Rappaport K, Cisternas M et al. (2005) Differences in patient and procedure characteristics and hospital resource use in primary and revision total joint arthroplasty. *Journal of Arthroplasty* **20**: 17–25.

Brosseau L, Davis J, Drouin H, Milne S, Noel M, Robinson VA et al. (2006) Continuous passive motion following total knee arthroplasty. *Cochrane Library* **1** http://www.thecochranelibrary.comCD004260.

Brunenberg DE, van Steyn MJ, Sluimer JC, Bekebrede LL, Bulstra SK, Joore MA (2005) Joint recovery programme versus usual care: an economic evaluation of a clinical pathway for joint replacement surgery. *Medical Care* **43**: 1018–1026.

Buechel FF Sr (2002) Long-term follow-up after mobile-bearing total knee replacement. *Clinical Orthopaedics and Related Research* **404**: 40–50.

Denis M, Moffet H, Caron F, Ouellet D, Paquet J, Nolet L (2006) Effectiveness of continuous passive motion and conventional physical therapy after total knee arthroplasty: a randomized clinical trial. *Physical Therapy* **86**: 174–185.

Dennis DA, Komistek RD, Hoff WA, Gabriel SM (1996) In vivo knee kinematics derived using an inverse perspective technique. *Clinical Orthopaedics and Related Research* **331**: 107–117.

Dennis DA, Komistek RD, Colwell CE Jr, Ranawat CS, Scott RD, Thornhill TS et al. (1998) In vivo anteroposterior femorotibial translation of total knee arthroplasty: a multicenter analysis. *Clinical Orthopaedics and Related Research* **356**: 47–57.

Diduch DR, Insall JN, Scott WN, Font-Rodriguez D (1997) Total knee replacement in young, active patients: long-term follow-up and functional outcome. *Journal of Bone and Joint Surgery* **79A**: 575–582.

Dixon T, Shaw M, Ebrahim S, Dieppe P (2004) Trends in hip and knee joint replacement: socioeconomic inequalities and projections of need. *Annals of Rheumatic Diseases* **63**: 825–830.

Dowsey M, Kilgour ML, Santamaria NM, Choong FM (1998) Clinical pathways in hip and knee arthroplasty: a prospective randomized controlled study. *Medical Journal of Australia* **170**: 59–62.

Duffy GP, Berry DJ, Rand JA (1998) Cement versus cementless fixation in total knee arthroplasty. *Clinical Orthopaedics and Related Research* **356**: 66–72.

Ethgen O, Bruyere O, Richy F, Dardennes C, Reginster JY (2004) Health-related quality of life in total hip and total knee arthroplasty. *Journal of Bone and Joint Surgery* **86A**: 963–971.

Erler K, Anders C, Fehlberg G, Neumann U, Brucker L, Scholle HC (2001) Measurements of results of a special hydrotherapy during in-patient rehabilitation after implantation of a total knee arthroplasty. *Zeitschrift für Orthopädie und ihre Grenzgebiete* **139**: 352–358.

Fortin PR, Penrod JR, Clarke AE, St-Pierre Y, Joseph L, Belisle P et al. (2002) Timing of total joint replacement affects clinical outcomes among patients with osteoarthritis of the hip or knee. *Arthritis & Rheumatism* **46**: 3327–3330.

Fortin PR, Clarke AE, Liang JL, Tanzer M, Ferland D et al. (1999) Outcomes of total hip and knee replacement: preoperative functional status predicts outcomes at six months after surgery. *Arthritis & Rheumatism* **42**: 1722–1728.

Fransen M, Crosbie J, Edmonds J (2003) Isometric muscle force measurement for clinicians treating patients with osteoarthritis of the knee. *Arthritis & Rheumatism* **49**: 29–35.

Fransen M, Crosbie J, Edmonds J (2001) Physical therapy is effective for patients with osteoarthritis of the knee: a randomized controlled clinical trial. *Journal of Rheumatology* **28**: 156–164.

Fransen M, McConnell S, Bell M (2002) Therapeutic exercise for people with osteoarthritis of the hip or knee: a systematic review. *Journal of Rheumatology* **29**: 1737–1745.

Frost H, Lamb SE, Robertson S (2002) A randomized controlled trial of exercise to improve mobility and function after elective knee arthroplasty. Feasibility, results and methodological difficulties. *Clinical Rehabilitation* **16**: 200–209.

Ganz SB, Benick RA (2004) A comparison of functional recovery following unilateral and bilateral total knee arthroplasty. *Topics in Geriatric Medicine* **20**: 310.

Ganz SB, Ranawat CS (2004) Efficacy of formal knee flexion exercises on the achievement of functional milestones following total knee arthroplasty. *Topics in Geriatric Medicine* **20**: 311.

Gibbons CER, Solan MC, Ricketts DM, Patterson M (2001) Cryotherapy compared with Robert Jones bandage after total knee replacement: a prospective randomized trial. *International Orthopaedics* **25**: 250–252.

Gotlin RS, Hershkowitz S, Juris PM, Gonzalez EG, Scott WN, Insall JN (1994) Electrical stimulation effect on extensor lag and length of hospital stay after knee arthroplasty. *Archives of Physical Medicine and Rehabilitation* **75**: 957–959.

Gur H, Cakin N, Akova B, Okay E, Kucukoglu S (2002) Concentric versus combined concentric-eccentric isokinetic training: effects on functional capacity and symptoms in patients with osteoarthrosis of the knee. *Archives of Physical Medicine and Rehabilitation* **83**: 308–316.

Hamdan TA, Al-Essa KA (2003) Manipulation under anaesthesia for the treatment of frozen shoulder. *International Orthopaedics* **27**: 107–109.

Haug J, Wood LT (1988) Efficacy of neuromuscular stimulation of the quadriceps femoris during continuous passive motion following total knee arthroplasty. *Archives of Physical Medicine and Rehabilitation* **69**: 423–424.

Healy WL, Seidman J, Pfeifer BA, Brown DG (1994) Cold compressive dressings after total knee arthroplasty. *Clinical Orthopaedics and Related Research* **299**: 143–146.

Heck D, Robinson R, Partridge C, Lubitz R, Freund D (1998) Patient outcomes after knee replacement. *Clinical Orthopaedics and Related Research* **356**: 93–110.

Herbert, R, Jamtvedt, G, Mead, J, Hagen, KB (2005) *Practical Evidence-Based Physiotherapy* Edinburgh: Elsevier.

Hirsch HS, Lotke PA, Morrison LD (1994) The posterior cruciate ligament in total knee surgery. Save, sacrifice, or substitute? *Clinical Orthopaedics and Related Research* **309**: 64–68.

Holt G, Dennis D (2003) The role of patella resurfacing in total knee arthroplasty. *Clinical Orthopaedics and Related Research* **1**: 76–83.

Insall JN, Ranawat CS, Scott WN, Walker P (1976) Total condylar knee replacement: preliminary report. *Clinical Orthopaedics and Related Research* **120**: 149–154.

Ivey M, Johnston RV, Uchida T (1994) Cryotherapy for postoperative pain relief following knee arthroplasty. *Journal of Arthroplasty* **9**: 285–290.

Jain NB, Guller U, Pietrobon R, Bond TK, Higgins LD (2005) Comorbidities increase complication rates in patients having arthroplasty. *Clinical Orthopaedics and Related Research* **435**: 232–238.

Jones DL, Westby MD, Greidanus N, Johanson NA, Krebs DE, Robbins L et al. (2005) Update on hip and knee arthroplasty: current state of evidence. *Arthritis & Rheumatism* **53**: 772–780.

Kajino A, Yoshino S, Kameyama S, Kohda M, Nagashima S (1997) Comparison of results of bilateral total knee arthroplasty with and without patellar replacement for rheumatoid arthritis: a follow-up note. *Journal of Bone and Joint Surgery* **79A**: 570.

Keblish PA, Varma AK, Greenwald AS (1994) Patellar resurfacing or retention in total knee arthroplasty: a prospective study of patients with bilateral replacements. *Journal of Bone and Joint Surgery* **76B**: 930.

Kennedy DM, Stratford PW, Wessel J, Gollish JD, Penney D (2005) Assessing stability and change of four performance measures: a longitudinal study evaluating outcome following total hip and knee arthroplasty. *BMC Musculoskeletal Disorders* **6**(3).

Kim BS, Reitman RD, Schai PA, Scott RD (1999) Selective patellar nonresurfacing in total knee arthroplasty: 10 year results. *Clinical Orthopaedics and Related Research* **367**: 81.

Kramer JF, Speechley M, Bourne R, Rorabeck C, Vaz M. (2003) Comparison of clinic- and home-based rehabilitation programs after total knee arthroplasty. *Clinical Orthopaedics and Related Research* **410**: 225–234.

Lachiewicz PF (2000) The role of continuous passive motion after total knee arthroplasty. *Clinical Orthopaedics and Related Research* **1**: 144–150.

Lamb SE, Frost H (2003) Recovery of mobility after knee arthroplasty. Expected rates and influencing factors. *Journal of Arthroplasty* **18**: 575–581.

Lingard EA, Bervan S, Katz JN, Kinemax Outcomes Group (2000) Management and care of patients undergoing total knee arthroplasty: variations across different health care settings. *Arthritis Care and Research* **13**: 129–136.

Lorentzen JS, Petersen MM, Brot C, Madsen OR (1999) Early changes in muscle strength after total knee arthroplasty. *Acta Orthopedica Scandinavica* **70**: 176–179.

March LM, Cross M, Tribe KL, Lapsley HM, Courtenay BG, Cross MJ et al. (2004) Two knees or not two knees? Patient costs and outcomes following bilateral and unilateral total knee joint replacement surgery for OA. *Osteoarthritis Cartilage* **12**: 400–408.

March LM, Cross MJ, Lapsley H, Brnabic AJ, Tribe KL, Bachmeier CJM et al. (1999) Outcomes after hip or knee replacement surgery for osteoarthritis. *Medical Journal of Australia* **171**: 235–238.

Martin SS, Spindler KP, Tarter JW, Detwiler KB (2002) Does cryotherapy affect intraarticular temperature after knee arthroscopy? *Clinical Orthopaedics and Related Research* **1**: 184–189.

McAuley JP, Harrer MF, Ammeen D, Engh GA (2002) Outcome of knee arthroplasty in patients with poor pre-operative range of motion. *Clinical Orthopedics and Related Research* **404**: 203–207.

McDonald S, Hetrick S, Green S (2004) Pre-operative education for hip or knee replacement. *Cochrane Library* **1** http://www.thecochranelibrary.com CD003526.

Milne S, Brosseau L, Robinson V, Noel MJ, Davis J, Drouin H et al. (2003) Continuous passive motion following total knee arthroplasty. *Cochrane Library* **2** http://www.thecochranelibrary.com CD004260.

Mitchell C, Walker J, Walters S, Morgan AB, Binns T, Mathers N (2005) Costs and effectiveness or pre- and post-operative physiotherapy for total knee replacement: randomized controlled trial. *Journal of Evaluation in Clinical Practice* **11**: 283–292.

Mizner RL, Stevens JE, Snyder-Mackler L (2003) Voluntary activation and decreased force production of the quadriceps femoris muscle after total knee arthroplasty. *Physical Therapy* **83**: 359–365.

Mizner RL, Petterson SC, Stevens JE, Axe MJ, Snyder-Mackler L (2005) Preoperative quadriceps strength predicts functional ability one year after total knee arthroplasty. *Journal of Rheumatology* **32**: 1533–1539.

Moffet H, Collet J-P, Shapiro SH, Paradis G, Marquis F, Roy L (2004) Effectiveness of intensive rehabilitation on functional ability and quality of life after first total knee arthroplasty: a single-blind randomised controlled trial. *Archives of Physical Medicine and Rehabilitation* **85**: 546–556.

Munin MC, Rudy TE, Glynn NW, Crossett LS, Rubash HE (1998) Early inpatient rehabilitation after elective hip or knee arthroplasty. *Journal of the American Medical Association* **279**: 847–852.

Australian Bureau of Statistics (1995) *National Health Survey SF-36 Population Norms* Canberra: Australian Bureau of Statistics, Cat. No. 4399.0.

Naylor JM, Ireland JE, Mohammed M (2005) *Outcomes Following Primary THR and TKR: part 1 – acute and short-term outcomes* Sydney: Whitlam Joint Replacement Centre, SSWAHS.

Naylor JM, Fransen M, Ireland JE, Winstanley J (2006a) *Outcomes Following Primary THR and TKR: part 2 – longer-term outcomes.* Sydney: Whitlam Joint Replacement Centre, SSWAHS.

Naylor JM, Harmer AR, Fransen M, Crosbie J, Innes L (2006b) The status of physiotherapy rehabilitation following total knee replacement in Australia. *Physiotherapy Research International* **11**: 35–47.

NHMRC and ASBT (2001) Clinical practice guideline on the use of blood components. http://www.nhmrc.health.gov.au.

Oldemeadow L, McBurney H, Robertson V (2001) Hospital stay and discharge outcomes after knee arthroplasty. *Journal of Quality and Clinical Practice* **21**: 56–60.

Ouellet D, Moffet H (2002) Locomotor deficits before and two months after knee arthroplasty. *Arthritis & Rheumatism* **47**: 484–493.

Paul RG, Bailey AJ (1996) Glycation of collagen: the basis of its central role in the late complications of ageing and diabetes. *International Journal of Biochemistry and Cell Biology* **28**: 1297–1310.

Pearson S, Moraw I, Maddern GJ (2000) Clinical pathway management of total knee arthroplasty: a retrospective comparative study. *ANZ Journal of Surgery* **70**: 351–354.

Petterson SC, Mizner RL, Snyder-Mackler L (2003) Factors that influence six-minute walk in individuals after total knee arthroplasty. *Journal of Geriatric Physical Therapy* **26**: 50.

Pierson J, Earles D, Wood K (2003) Brake response time after total knee arthroplasty. *Journal of Arthroplasty* **18**: 840–843.

Rajan RA, Pack Y, Jackson H, Gillies C, Asirvatham R (2004) No need for outpatient physiotherapy following total knee arthroplasty: a randomized trial of 120 patients. *Acta Orthopedica Scandinavica* **75**(1): 71–73.

Rand JA (1996) Posterior cruciate retaining total knee arthroplasty. In: Morrey BF (ed.) *Reconstructive Surgery of the Joints* Volume 2 (2 edn) New York: Churchill Livingstone, pp. 1401–1408.

Ritter MA, Berend ME, Meding JB, Keating EM, Faris PM, Crites BM (2001) Long-term follow-up of anatomic graduated components posterior cruciate-retaining total knee replacement. *Clinical Orthopedics* **388**: 51–57.

Rodriguez JA, Bhende H, Ranawat CS (2001) Total condylar knee replacement: a 20-year follow-up study. *Clinical Orthopaedics and Related Research* **388**: 10–17.

Roos E (2003) Effectiveness and practice variation of rehabilitation after joint replacement. *Current Opinion in Rheumatology* **15**: 160–162.

Rossi MD, Hasson S (2004) Lower-limb force production in individuals after unilateral total knee arthroplasty. *Archives of Physical Medicine and Rehabilitation* **85**: 1279–1284.

Salmon P, Hall G, Peerbhoy D, Shenkin A, Parker C (2001) Recovery from hip and knee arthroplasty: patients' perspective on pain, function, quality of life, and well-being up to 6 months postoperatively. *Archives of Physical Medicine and Rehabilitation* **82**: 360–366.

Scarcella JB, Cohn BT (1995) The effect of cold therapy on the postoperative course of total hip and knee arthroplasty patients. *American Journal of Orthopedics* November: 847–852.

Segal L, Day SE, Chapman AB, Osborne RH (2004) Can we reduce disease burden from osteoarthritis? An evidence-based priority-setting model. *Medical Journal of Australia* **180**: 11S–17S.

Shields RK, Enloe LJ, Leo KC (1999) Health related quality of life in patients with total hip or knee replacement. *Archives of Physical Medicine and Rehabilitation* **80**: 572–579.

Shumway-Cook A, Brauer S, Woollacott M. (2000) Predicting the probability for falls in community-dwelling older adults using the timed up and go test. *Physical Therapy* **80**: 897–903.

Sigal RJ, Kenny GP, Wasserman DH, Castaneda-Sceppa C (2004) Physical activity/exercise and type 2 diabetes. *Diabetes Care* **27**: 2518–2539.

Skinner J, Weinstein JN, Sporer SM, Wennberg JE (2003) Racial, ethnic, and geographic disparities in rates of knee arthroplasty among Medicare patients. *New England Journal of Medicine* **349**: 1350–1359.

Smith J, Stevens J, Taylor M, Tibbey J (2002) A randomised controlled trial comparing compression bandaging and cold therapy in postoperative total knee replacement surgery. *Orthopaedic Nursing* **21**: 61–66.

Steffen T, Hacker TA, Mollinger L (2002) Age- and gender-related test performance in community-dwelling elderly people: six-minute walk test, berg balance scale, timed up & go test, and gait speeds. *Physical Therapy* **82**: 128–137.

Stratford PW, Kennedy D, Pagura SM, Gollish JD (2003) The relationship between self-report and performance-related measures: questioning the content validity of timed tests. *Arthritis & Rheumatism* **49**: 535–540.

Straw R, Kulkarni S, Attfield S, Wilton TJ (2003) Posterior cruciate ligament at total knee replacement. Essential, beneficial or hindrance? *Journal of Bone and Joint Surgery* **85B**: 671–674.

van Essen GJ, Chipcase LS, O'Connor D, Krishnan J (1998) Primary total knee replacement: short-term outcomes in an Australian population. *Journal of Quality in Clinical Practice* **18**: 135–142.

Walsh M, Woodhouse LJ, Thomas SG, Finch E (1998) Physical impairments and functional limitations: a comparison of individuals 1 year after total knee arthroplasty with control subjects. *Physical Therapy* **78**: 248–258.

Wang A, Hall S, Gilbey H, Ackland T (1997) Patient variability and the design of clinical pathways after total hip replacement surgery. *Journal of Quality in Clinical Practice* **17**: 123–129.

Webb JM, Williams D, Ivory JP, Day S, Williamson DM (1998) The use of cold compression dressings after total knee replacement: a randomised controlled trial. *Orthopedics* **21**: 59–61.

Whiteside LA (1995) Ligament balancing and bone grafting in total knee replacement of the varus knee. *Orthopedics* **18**: 117–122.

Whiteside LA (1999) Selective ligament release in total knee arthroplasty of the knee in valgus. *Clinical Orthopedics and Related Research* **367**: 130–140.

Whiteside LA. (2005) Assess and release the tight ligament. In: Bellemans J, Ries MD, Victor J (eds) *Total Knee Arthroplasty: a guide to get better performance*, pp. 170–176.

Wood DJ, Smith AJ, Collopy D, White B, Brankov B, Bulsara MK (2002) Patellar resurfacing in total knee arthroplasty: a prospective, randomized trial. *Journal of Bone and Joint Surgery* **84A**: 187–193.

Index